INSECTS OF LATIN AMERICA

A PHOTOGRAPHIC GUIDE

Text and photographs by Stephen Cresswell

Second edition

TROX PRESS 2020

Buckhannon, West Virginia

Second edition

ISBN 978-0-578-80437-8 (Paperback edition)

Cover design by Fiaz Ahmed

Printing 1a

CONTENTS

Introduction 5

1 Order Odonata: Dragonflies and Damselflies 7

2 Order Orthoptera: Grasshoppers, Crickets, and Katydids 25

3 Order Phasmatodea: Stick Insects 41

4 Order Mantodea: Mantises 49

5 Order Blattodea: Cockroaches and Termites 63

6 Order Hemiptera: True Bugs, Cicadas, Hoppers, and Aphids 71

7 Orders Megaloptera and Neuroptera: Dobsonflies, Lacewings, Owlflies, Mantisflies and Antlions 139

8 Order Coleoptera: Beetles 147

9 Order Diptera: Flies 207

10 Order Trichoptera: Caddisflies 235

11 Order Lepidoptera: Moths and Butterflies 241

12 Order Hymenoptera: Sawflies, Wasps, Bees, and Ants 317

13 Other Insect Orders: Bristletails, Mayflies, Stoneflies, Earwigs, Webspinners, Bark Lice, Thrips 355

Books for Further Reading 360

Index 363

Figure 1 Four fulgorid planthoppers, *Enchophora sanguinea*, feeding on sap on a tree trunk, Cartago, Costa Rica

INTRODUCTION

This book is a photographic guide to insects that live in the Neotropics, the biogeographic region located south of the United States and extending south to Tierra del Fuego.

The photos that make up this volume were taken in the Neotropics between 2008 and 2020. Every mainland country of Latin America except Venezuela is represented. Photos are included from Caribbean island nations and dependencies, including Haiti, the Dominican Republic, Jamaica, Puerto Rico, the U.S. Virgin Islands, Martinique, Aruba, and Trinidad and Tobago. In nearly all cases experts have examined the photographs and made an identification—ideally to species, but in some cases to genus, tribe, or subfamily. In most cases the experts examined two or three large photos taken from different angles. The experts also had the date of the photo, the habitat type, the location including elevation, and in many instances other information such as the food plant, or the insect's length.

Like most field guides of various kinds, this book has identifications based on photographs and not on museum specimens. Some readers may object that pinned specimens are needed for a reliable ID. Well, even with a pinned specimen not all experts will agree on a species ID. Still, when a photograph's caption in this book says, for example, "Plesiochrysa brasiliensis" it is better to read that as "The insect in the photo is consistent with *Plesiochrysa brasiliensis*" rather than "without any doubt, the insect is *Plesiochrysa brasiliensis*." With at least two million insect species in the world, described and undescribed, this seems the most reasonable approach.

Where the expert identifier expressed a lower level of confidence in the ID we have carried that concern over into the listings, for example using a "cf." between the genus and species names. Use of the cf. suggests that the genus is likely correct, but that one should "compare with" the species in question, which is, perhaps, the most likely ID for the insect pictured. If the genus ID is likely but with some room for doubt we have used an "equals approximately" symbol before the genus name, for example: ≈ Platyphora sp.

Warmest thanks to the many scientists who shared their time and their expertise to identify photos. Without them this book would have been impossible. Thanks to: John Abbott, Jay Abercrombie, Bob Acciavatti, Luiz Alexandre Campos, Jorge Almeida, Robert S. Anderson, John S. Ascher, Thomas H. Atkinson, Kyhl Austin, José Avendaño, Manuel Baena, Kevin Barber, Jérôme Barbut, Charles Bartlett, Rafael Barros, Sara Bazyar, Paul Bedell, Chuck Bellamy, Vassili Belov, Cesar João Benetti, Øistein Berg, Matt Bertone, Paul Beuk, Larry Bezark, Wes Bicha, Gabriel Biffi, Roger Blahnik, Lech Borowiec, Craig Brabant, Harry Brailovsky, Gavin Broad, Paul Brock, John W. Brown, Richard L. Brown, Sharon Brown, Matthias Buck, Gerry Bunker, John F. Burger, Kelsey J.R.P. Byers, Oscar J. Cadena-Castañeda, Andy Calderwood, Lucas Denadai de Campos, Claudio J.B. Carvalho, Carlos S. Carbonell, Philip Careless, Bob Carlson , John Carlson, James M. Carpenter, John F. Carr, Scott P. Carroll, Leopoldo Castro, Pierfilippo Cerretti, Donald S. Chandler, Shawn Clark, Rebecca Clayton, Andrew Cline, Joe Cicero, Anthony Cognato, Nancy Collins, Steve Collins, Oskar Conle, Glen Corbiere, Ben Coulter, Greg Courtney, Mauro Daccordi, Robert Davidson, Paul Davis, Pablo M. Dellapé, Aristeo Cuahtemoc Deloya Lopez, Chris Dietrich, Dmitry Dmitriev, Greg Dodge, Stephanie A. Dole, Jason J. Dombroskie, Nick Donnelly, Peter Duelli, Steen Dupont, Libor Dvořák, Dave Edmonds, Joe Eger, Carlos José Einicker Lamas, Randy Emmitt, Terry Erwin, Flávia Esteves, Dominic Evangelista, Omar Fahmy, Rafaela Falaschi, Nick Fensler, David J. Ferguson, Marcos Fianco, Andressa Figueiredo de Oliveira, Eric Fisher, Wills Flowers, Dimitri Forero, Rafael Freitas-Silva, Paul Freytag, David G. Furth, Steve Gaimari, Rosser W. Garrison, Héctor Jaime Gasca Alvarez, Michael Geiser, Yann Gobeil, Gil Felipe Gonçalves Miranda, Lloyd Gonzales, Guillermo González, Ranulfo González-Obando, Jacob Gorneau, Andrej Gorochov, Harold Grau, Igor Grichanov, Grey Gustafson, Fabian Haas, Dennis Haines, Robert Hamilton, Paul Hanson, Martin Hauser, James Hayden, Sam W. Heads, Ernst Heiss, Charles S. Henry, Thomas J. Henry, Daniel Herbin, Henry Hespenheide, Auke J. Hielkema, Chris Hill, Ross Hill, Dominik Hofer, Kevin Hoffman, Kamil Holy, Ralph Holzenthal, Werner Holzinger, Nick Homziak, Michael Huben, Tony Irwin, Michael A. Ivie, Mary Liz Jameson, Daniel Janzen, Alan Jeon, Paul Johnson, Lucas Kaminski, S. V. Kazantsev, Gabor Keresztes, Won Gun Kim, Lynn Kimsey, Lawrence Kirkendall, Victor Kolyada, Takumasa Kondo, Valery Korneyev, Michel Laguerre, Zachary Lahey, Ed Lam, Bernard Landry, Alfredo Lanuza-Garay, Stephane Lebrun, Sonia Maria Lopes Fraga, Augusto Loureiro Henriques, William Mackay, Ted C. MacRae, Jean-Michel Maes, Chris Majka, Dr. José Antonio Marin Fernandes, Pablo Marino, Paul Marsh, Blaine Mathison, Kátia Matiotti, Bill Mauffray, Peter May, Alfredo Mayorga, Stuart H. McKamey, J.E. McPherson, Ximo

Mengual, Jose Ricardo M. Mermudes, Peter W. Messer, Jan Metlevski, Julio Micheli, Lukasz Mielczarek, Laura T. Miller, Ed Mockford, Marina Monjardim, Miguel Monné, Irina Morales, Odette Morvan, William L. Murphy, Declan A. Murray, Stefan Naglis, Elynton Nascimento, Robert Nash, Piotr Naskrecki, Eugenio Nearns, Jhon César Neita-Moreno, Alex Netherton, Allen Norrbom, Ezequiel Núñez Bustos, Charles W. O'Brien, Lois B. O'Brien, Jim O'Hara, Michael Ohl, John D. Oswald, Thomas Pape, Abdiel Park-Martin, Bob Patterson, Dennis Paulson, Stewart Peck, Gerard Pennards, Norman Penny, Matthew Pintar, Gabriela Pirani, Walter Pfliegler, James Pitts, David Plotkin, Jerry A. Powell, Jens Prena, Wojciech Pulawski, Donald L.J. Quicke, José A. Rafael, Herschel Raney, Chris Raper, Claus Rasmussen, Brett C. Ratcliffe, John Rawlins, Menno Reemer, Bill Reynolds, Brady Richards, Julio Rivera, Tony Robillard, Roger Rohrbeck, Garrie Rouse, C.H.F. Rowell, Dave Ruiter, Björn Rulik, Ryan St. Laurent, Albino Sakakibara, Allen F. Sanborn, Diego de Santana Souza, Davide Sassi, Carl Schaefer, Christian Schmid-Egger, Chris Schmidt, Michael Schmitt, Jack Schuster, Michael D. Schwartz, Paulo Schwirkowski, Lukas Sekerka, Alessandra Sforzi, Igor Shamshev, Michael Sharkey, Scott R. Shaw, Floyd Shockley, Andrew Short, Gervasio Silva Carvalho, Tatiana Sepúlveda, Luiz Silveira, Rebecca Simmons, Bradley Sinclair, Paul E. Skelley, Jeff Skevington, Dave Smith, Gordon C. Snelling, Gilberto Soares Albuquerque, Ángel Solís, John Smit, Charles L. Staines, Lionel Stange, Warren Steiner, Eckart Stolle, D.R. Swanson, Merrill Sweet, Karol Szawaryn, William Taft, Daniela M. Takiya, Catherine Ann Tauber, Camille Thirion, Donald B. Thomas, Michael C. Thomas, Tony Thomas, Vinton Thompson, James C. Trager, Pablo Valero, Ruud van der Weele, Joke Van Erkelens, Gergely Várkonyi, Richard Vernier, Nikita Vikhrev, Cecilia Waichert, George Waldren, Susan Weller, Jeff Webb, Richard L. Westcott, Terry Wheeler, Terry Whitworth, Christopher M. Wilson, Michael Wilson, Stephen Wilson, Donald Windsor, Harmut Wisch, Monty Wood, Norm Woodley, Andrzej J. Woznica, Douglas Yanega, Matt Yoder, Chen Young, Daniel K. Young, Diane Young, James Zahniser, Theo Zeegers, Edison Zefa, Álvaro Zúñiga-Reinoso, and Bob Zuparko.

As this volume goes to press the world's economy is partly shut down by COVID-19 and travel between countries has come to a standstill. Air pollution by the most industrialized countries has led to climate change, seen in more severe droughts in some areas and destructive flooding in others. Sea levels are rising as polar ice caps and glaciers melt. Meanwhile habitat destruction to feed the world's growing population has led to extinctions and fears of more extinctions. As many writers have noted, it is imperative that we study this amazing level of diversity of life on earth while we still can. Of course, it is equally important to slow and eventually stop climate change, and to learn how to feed the world's people without wholesale destruction of pristine areas.

SAMPLE ENTRY

Acrididae: Ommatolampidinae

Abacris flavolineata • Short-horned Grasshopper

Playa Los Angeles, Magdalena, Colombia

Line one gives the family and subfamily. Line two gives the scientific name (usually a binomial) and a common name. The third line gives the place where the photo was taken.

Note that in some cases the common name refers to the species, but in other cases it refers to the genus, subfamily, or family. Where a common name was lacking, in some cases we translated the Latin or Greek scientific name to provide a common name.

In giving the place where the photo was taken, English spellings are used for countries, but the local languages are used for towns and provinces. So, for example, an entry might read, "Lago Azul, Panamá Oeste, Panama."

▶ **Entries should be read column by column rather than row by row. Doing this will keep closely related insects together.**

Corrections to the identifications, or comments on them, are most appreciated, and will help in the preparing of future editions.

1

ORDER ODONATA

Dragonflies and Damselflies

Many a naturalist has sat by a pond, transfixed by the beauty of the dragonflies that are flying and perching, hunting and mating, displaying diaphanous wings and bodies brilliantly colored and spotted and striped. Every now and then, two flying males meet to fight over territorial boundaries, and the quiet is broken by the sound of clashing wings.

Although dragonflies and damselflies make up a much smaller group than moths or beetles or wasps, they excite a lot of interest in insect watchers because of their beauty, their interesting life histories, and their astonishing flying abilities.

The world's first flying creatures were the dragonflies and their immediate ancestors. Unlike most insect groups that evolved later, members of the order Odonata have large wings that do not fold up when at rest. Unlike other insect groups, odonates have their wings attached directly to the muscles that power them. In other insect orders the wings move as a result of a flexing of the body walls. Dragonfly wings beat much more slowly than the wings of most other insect groups, but nevertheless dragonflies are superb flyers.

These airborne dragons can fly slow or fast, can hover, and do loop-the-loops. Damselflies fly more slowly, but are skilled at moving in and out of thick stands of herbaceous plants, quickly snatching their insect prey from on leaves.

Dragonflies have wings that are large and many-veined; the veins form a net-like pattern. Damselflies also have many-veined wings, often stalked. Both groups commonly have a small spot along the front edge where the wing may be thickened; this is called a pterostigma, and its appearance can be helpful in making identifications. Both groups have tiny antennae. It is common in both groups to have males and females of different colors, for example the male blue and the female green. For this reason we have marked all the photos in this chapter with a male or female symbol (♂ or ♀).

Order Odonata is divided into two suborders, both widespread in the Neotropics.

One is the Anisoptera, the dragonflies. Though there are some exceptions, generally dragonflies are large, robust insects that are fast fliers. At rest they tend to hold their wings out to the sides, or else pointing straight up. Usually their compound eyes meet on the top of the head, or at least are very close to each other there.

By contrast, members of the suborder Zygoptera, the damselflies, tend to be long and slender and overall smaller. Damselfly eyes are widely separated on top of their heads. Damselflies are slower in flight, and generally less acrobatic when flying. At rest they usually hold their four wings together above the abdomen and parallel to it. A smaller number of damselfly groups hold their wings partly open so that they meet the body diagonally.

The Neotropics are home to about 820 species of dragonfly and 930 species of damselfly, for a total of 1,750 described odonate species. Other species await description.

Reproduction. In many species of odonate, males and females find each other by sight and begin mating with few preliminaries. In other species, particularly among the damselflies, there may be elaborate courtship displays, including special flight patterns, displays of colorful wings or legs, and even a "standing on the water" display in which a male jewelwing floats by the perched female. Some male odonates maintain a territory, excluding other males and attempting to mate with any female that enters their area. Males in other groups, such as darners, do not maintain territories, but seek females wherever they may find them.

People are often fascinated to see mating odonates forming a circular or heart-shaped hoop, either at rest or flying through the air. The details of odonate mating are indeed surprising. Males have the normal complement of sex organs at the tip of the abdomen, but also have an auxiliary penis on the ventral (bottom) side of their abdomen near where it joins the thorax. Males deposit sperm from the tip of their abdomen onto the auxiliary penis, then hook up with a female by holding her head or "neck" with claspers located on his abdominal apex. The female then loops her abdominal tip forward, attaching it to the male's auxilliary penis.

Oviposition. All odonates lay their eggs in water—or in the case of a few groups, in the very damp soil of a seep. Different species of odonate have differing methods of depositing their eggs. Some female dragonflies deposit eggs one by one, others as clumps, and others as a long gelatinous string. (The string of eggs often snag on vegation that will provide hiding places for the nymphs when they hatch.) Some dragonflies release their eggs from high above the water, while others dip their abdomen repeatedly in the water to release the eggs. The latter method is dangerous, of course, each dip leaving the female vulnerable to predation by fish and other aquatic creatures.

Other egg-laden females, including all damselflies and some dragonflies, possess an ovipositor that bears two serrated blades used to cut slits into plant stems or other plant parts. The elongated egg is then laid into the prepared slit. Odonates that place their eggs within the protective confines of a plant typically produce and lay fewer eggs than their broadcasting relatives.

As the female oviposits, males of some groups engage in mate guarding, aiming to prevent other males from mating with the female until the eggs are laid. Sometimes this guarding is done from a short distance away, while in other cases the pair remain attached ("in tandem") with the tip of the male's abdomen attached to the female's head. In the last 40 years researchers have discovered that in some odonate species the male uses his penis as a tool to remove the sperm of other males from the female's body in order to make it likely that only his sperm will fertilize the eggs.

Figure 2 A female *Acanthagrion sp.* damselfly deposits her eggs while above her a male is ready to repel any rival males that may arrive. Uruguay.

Larvae. Odonates are creatures of incomplete metamorphosis (hemimetabolism), lacking a pupal stage. Larvae are aquatic, while adults are terrestrial and aerial. In incomplete metamorphosis the immatures look somewhat like small adults, though lacking developed wings. The larvae in insects with incomplete metamorphosis are often called nymphs, but larvae is equally correct.

Damselfly larva are long and slender, like the adults. In many species the larvae cling to plants, and may escape the notice of predators because they resemble plant stems. Damselfly larvae are relatively feeble swimmers, and their slender bodies allow them to move through the water with less water resistance than a thicker body would permit. Damselfly larvae have external gills located at the hindmost part of their abdomens. Some damselfly species use these gills as paddles for locomotion.

Dragonfly larvae tend to be large and robust, even plump. They have a remarkable method of rapid locomotion that means a streamlined body is not needed. These larvae are able to expel water very rapidly from their rectum, causing them to shoot forward at great speed. Their rectum is also notable for containing the gills. Since respiration is taking place in the rectum, the feces are enveloped in a membrane so as not to dirty the respiration process.

Both dragonfly and damselfly larvae possess extraordinary labial palps, spoon-shaped structures that at rest are "cocked" and stockpiling kinetic energy. When near prey, the larva can release the palps, which shoot forward to clamp and hold the target species. Like the adults, larvae feed primarily on other arthropods, but they are also among the few insects to successfully hunt and kill vertebrate prey (such as minnows or tadpoles).

Dragonfly and damselfly larvae vary in their habits and habitats. Some sprawl on the bottom of the stream or pond and rely on camouflage to avoid detection. Some live among the sheltering leaves of water plants. Some hide amidst detritus on the bottom, while others burrow in the mud or sand there.

Larvae are effective hunters and grow as they feed, necessitating molts of the outgrown exoskeleton as they progress toward adulthood. As Dennis Paulson notes, "The variation in development time is enormous." Some species that live in warm climates and develop in temporary pools may go from egg to adult in as little as twenty days. Others that live in cold streams with reduced numbers of prey species may take eight years to develop. These long development times for the immatures of some species stand in stark contrast to the adult lives of odonates, which typically last only about a month.

Figure 3 This larval sanddragon *(Progomphus sp.)* was dredged up from the pebbly stream bottom to have its photo taken. These larval dragonflies are burrowers. Panama.

Emergence. As the developmental period ends, the larva emerges from the water, often climbing a plant stem just before dawn. Most species then swallow air, causing the body to expand and split the larval cuticle. With great effort the odonate frees itself from the old cuticle, and as it does so the wings (which had been accordian-folded) unfurl. At this point the adult odonate is called a "teneral," meaning that it has a soft cuticle and a relatively pale color (often light green or tan). This is a dangerous period, as the insect is not protected by a hard cuticle or the ability to fly.

Soon fluids are pumped into the wings, and these expand to full-size. As the wings expand and the cuticle hardens, the adult takes its first flight and and begins a new life of aerial hunting and of mating.

Feeding. Odonates in flight pursue and capture other insects, imprisoning them in the "basket" formed by their six bristly legs. Their success as predators is founded on their superb flying skills, also by their extraordinary vision provided by up to 30,000 ommatidia (facets) in each of two compound eyes. Damselflies most often dine on small prey such as mosquitos, midges leafhoppers, and other diminutive insects. Dragonflies prey on these smaller insects too, but larger dragonflies also seek out horseflies, moths, butterflies, and other odonates.

Perils and defenses. Although odonates have impressive hunting skills, they are also regularly preyed upon, both as larvae and as adults. Birds are the most important consumers of adult odonates. Other insects, too, often dine on dragonflies and damselflies; these include mantises, social wasps, and robber flies. Spiders, fish, frogs, turtles, and lizards catch and eat larval and adult odonates too. Also, odonates eat odonates.

Figure 4 Larval spreadwing damselfly *(Archilestes grandis)*. Its third abdominal gill is barely visible because it is in a plane perpendicular to the camera lens. Costa Rica.

Odonates are also harmed by parasites and parasitoids. Some of the most common odonate parasites are flies in the genus *Forcipomyia*. These insects use bladelike mouthparts to pierce odonate wing veins and feed on the haemolymph (analagous to vertebrate blood). Mites, including water mites, are another common parasite of odonates. These mites, often red and globular, slice into the body of an adult odonate (often a soft-bodied teneral) and consume haemolymph and liquified tissue. On the other hand, some of the mites seen on adult dragonflies and damselflies are only "hitching a ride" to another location and do no real harm; this practice is called phoresy.

Adults dragonflies and damselflies rely on a flying escape as their most important defense. Dragonflies can outfly many predators, and a large odonate can thrash its way out of a spider web. Larval anisopterans use their "jet propulsion" to flee danger. Camouflage is less often an important defense for odonates, compared with other insects. The males have bright colors and markings to attract mates and are not likely to blend in with most substrates. Drab colored females do achieve some benefit from blending in with dead leaves and the like. Males of some species are able to change relatively quickly into drab coloration, especially at lower temperatures when the adult is not warm enough to take flight.

Figure 5 The exuvia (cast-off skin) of a *Brechmorhoga sp.* dragonfly larva that molted into adulthood. Such exuviae can help odonate experts monitor dragonfly distribution. Panama.

Figure 6 A teneral female damselfly, *Ischnura fluviatilis,* waits while her integument hardens and her wings fill with haemolymph. Also in this short period her color will darken. Uruguay.

Neotropical specialties. A number of Nearctic odonate genera are absent from the American tropics—*Amphiagrion* (red damsels) and *Cordulia* (American emeralds), for example. On the other hand, some interesting odonate taxa are found only in the Neotropics.

One such group is the helicopter damselflies. These are very large damsels, with a prominent spot of color on the wing tips. They are found even in the darkest rainforests and seem almost ghostly as they move slowly in an unusual whirring flight. Morphologically the family is noted for having multi-celled pterostigmata, and behaviorally for their unusual prey specialization: they pluck spiders from webs. Also interesting is their developmental history. Helicopter damselflies spend their larval life in water in tree holes and in the "water tanks" of bromeliads.

Other interesting groups found only in the Neotropics are the shadowdamsels and threadtails. (One species of shadowdamsel is found, rarely, in the far south of Arizona.) These are thin-bodied damselflies that live in dense tropical forest. Shadowdamsels make up family Platygastridae. The threadtails were previously placed in their own family, but now are considered part of the large pond damsels family Coenagrionidae.

Aeshnidae: Aeshninae

Gynacantha nervosa • Twilight Darner ♀

Mountain Pine Ridge Forest Reserve, Belize

Aeshnidae: Aeshninae

Triacanthagyna ditzleri • Ditzler's Darner ♂

Comarca Indígena Emberá-Wounaan, Panama

Aeshnidae: Aeshninae

Gynacantha membranalis • Dark-saddled Darner ♀

Tuis, Cartago, Costa Rica

Libellulidae: Libellulinae

Misagria sp. • Skimmer ♀

Along the Rupununi River, Guyana

Aeshnidae: Aeshninae

Neuraeschna harpya • Darner ♂

Reserva Particular do Cristalino, Mato Grosso, Brazil

Libellulidae: Brachydiplacinae

Anatya guttata • Common Blue-eye ♂

Gamboa, Colón, Panama

Libellulidae: Brachydiplacinae
Anatya guttata • Common Blue-eye ♀
Gamboa, Colón, Panama

Libellulidae: Brachydiplacinae
Micrathyria pseudeximia • Tropical Dasher
Montsinéry, Montsinéry-Tonnegrande, French Guiana

Libellulidae: Brachydiplacinae
Micrathyria aequalis • Spot-tailed Dasher ♂
Lago de Yojoa, Cortés, Honduras

Libellulidae: Brachydiplacinae
Oligoclada nr. monosticha • Leafsitter ♂
Novo Airão, Amazonas, Brazil

Libellulidae: Brachydiplacinae
Micrathyria hypodidyma • Austral Dasher
Nueva Palmira, Colonia, Uruguay

Libellulidae: Brachydiplacinae
Uracis imbuta • Tropical Woodskimmer ♂
Lagunas de Aruza, Darién, Panama

Libellulidae: Brachydiplacinae
Uracis ovipositrix • Spiketail Woodskimmer ♀
Reserva Particular do Cristalino, Mato Grosso, Brazil

Libellulidae: Libellulinae
Cannaphila insularis • Gray-waisted skimmer ♀
Parque Natural Metropolitano, Panamá, Panama

Libellulidae: Libellulinae
Cannaphila vibex • Blue-tailed Skimmer ♂
Altos del Maria, Coclé, Panama

Libellulidae: Palpopleurinae
Diastatops pullata • Red-saddled Blackwing ♂
Colakreek, Para, Suriname

Libellulidae: Trithemistinae
Brechmorhoga rapax • Amber-banded Clubskimmer ♂
Selva Negra, Matagalpa, Nicaragua

Libellulidae: Trithemistinae
Dythemis nigra • Blue-eyed Setwing ♀
Berg en Dal, Brokopondo, Suriname

Libellulidae: Trithemistinae
Dythemis sterilis • Brown Setwing ♂
Lago Bayano, Panamá province, Panama

Libellulidae: Sympetrinae
Erythrodiplax fusca • Red-faced Dragonlet ♂
La Ermita, Panamá Oeste, Panama

Libellulidae: Sympetrinae
Erythrodiplax attenuata • Sooty-winged Dragonlet ♀
Novo Airão, Amazonas, Brazil

Libellulidae: Sympetrinae
Erythrodiplax sp. • Dragonlet ♀
Berg en Dal, Brokopondo, Suriname

Libellulidae: Sympetrinae
Erythrodiplax fervida • Red-mantled Dragonlet ♂
Berg en Dal, Brokopondo, Suriname

Libellulidae: Sympetrinae
Erythrodiplax funerea • Black-winged Dragonlet ♂
Lagunas de Aruza, Darién, Panama

Libellulidae: Sympetrinae
Erythrodiplax nigricans ● Blackish Dragonlet ♂
Colonia del Sacramento, Colonia, Uruguay

Libellulidae: Sympetrinae
Erythemis plebeja ● Pin-tailed Pondhawk ♀
Along the Uruguay River, Colonia, Uruguay

Libellulidae: Sympetrinae
E. umbrata ● Band-winged Dragonlet andromorph ♀
Gamboa, Colón, Panama

Libellulidae: Sympetrinae
Erythemis vesiculosa ● Great Pondhawk ♂
Juan Hombrón, Coclé, Panama

Libellulidae: Sympetrinae
Erythemis peruviana ● Flame-tailed Pondhawk ♂
Gamboa, Colón, Panama

Libellulidae: Sympetrinae
Rhodopygia cardinalis • Cardinal Redskimmer ♂
Berg en Dal, Brokopondo, Suriname

Libellulidae: Libellulinae
Libellula herculea • Hercules Skimmer ♂
Altos del Maria, Coclé, Panama

Libellulidae: Libellulinae
Orthemis sp. • Antillean Red Skimmer ♂
Caño Hondo, Hato Mayor, Dominican Republic

Libellulidae: Libellulinae
Orthemis discolor • Carmine Skimmer ♂
La Suiza, Cartago, Costa Rica

Libellulidae: Palpopleurinae
Perithemis mooma • Pond Amberwing ♂
Montsinéry, Montsinéry-Tonnegrande, French Guiana

Libellulidae: Libellulinae
Orthemis nodiplaga • Wingspot Skimmer ♂
Nueva Palmira, Colonia, Uruguay

Libellulidae: Palpopleurinae
Perithemis mooma • Pond Amberwing ♀
Parque Nacional Pico Bonito, Atlántida, Honduras

Libellulidae: Trithemistinae
Macrothemis extensa • Attenuate Sylph ♀
Naranjo, Alajuela, Costa Rica

Calopterygidae: Hetaerininae
Hetaerina caja • River Rubyspot ♂
Metetí, Darién, Panama

Libellulidae: Trithemistinae
Macrothemis pseudimitans • White-tailed Sylph ♂
La Suiza, Cartago, Costa Rica

Calopterygidae: Hetaerininae
Hetaerina caja • River Rubyspot ♀
Metetí, Darién, Panama

Calopterygidae: Hetaerininae
Hetaerina aurora • Aurora Rubyspot ♂
Nanegalito, Pichincha, Ecuador

Calopterygidae: Hetaerininae
Hetaerina cruentata • Highland Rubyspot ♂
Selva Negra, Matagalpa, Nicaragua

Calopterygidae: Hetaerininae
Hetaerina fuscoguttata • Dot-winged Rubyspot ♂
Soberanía National Park, Colón, Panama

Coenagrionidae: Argiinae
Argia anceps • Cerulean Dancer ♂ ♀
La Suiza, Cartago, Costa Rica

Calopterygidae: Hetaerininae
Hetaerina laesa • Red-and-green Rubyspot ♀
Manú National Park, Madre de Dios, Peru

Coenagrionidae: Argiinae
Argia elongata • Thorn-tipped Dancer ♂
Selva Negra, Matagalpa, Nicaragua

Coenagrionidae: Argiinae
Argia adamsi • Varied Dancer ♂
Soberanía National Park, Colón, Panama

Coenagrionidae: Argiinae
Argia fissa • Azure Dancer ♂
Altos del Maria, Coclé, Panama

Coenagrionidae: Argiinae
Argia cf. fulgida • Garnet Dancer ♂
Playa Los Angeles, Magdalena, Colombia

Coenagrionidae: Argiinae
Argia oculata • Oculate Dancer ♂
Tuis, Cartago, Costa Rica

Coenagrionidae: Argiinae
Argia cf. medullaris • Sky-blue Dancer ♂ ♀
Nanegalito, Tandayapa Valley, Ecuador

Coenagrionidae: Argiinae
Argia oculata • Oculate Dancer ♀
Sansón, Darién, Panama

Coenagrionidae: Argiinae
Argia pulla • Purple Dancer ♂ ♀
La Suiza, Cartago, Costa Rica

Coenagrionidae: Argiinae
Argia oenea • Fiery-eyed Dancer ♂
La Ermita, Panamá Oeste, Panama

Coenagrionidae: Argiinae
Argia translata • Dusky Dancer ♂
Metetí, Darién, Panama

Coenagrionidae: Ischnurinae
Acanthagrion sp. • Wedgetail ♂
Colakreek, Para, Suriname

Coenagrionidae: Pseudagrioninae
Anisagrion allopterum • Middle American Pearlwing ♀
Selva Negra, Matagalpa, Nicaragua

Coenagrionidae: Ischnurinae
Acanthagrion trilobatum • Pacific Wedgetail ♂
Valle de Antón, Coclé, Panama

Coenagrionidae: Coenagrioninae
Neoerythromma cultellatum • Caribbean Yellowface ♂
Gamboa, Colón, Panama

Coenagrionidae: Ischnurinae
Cyanallagma bonariense • Buenos Aires Damsel ♂ ♀
Colonia del Sacramento, Colonia, Uruguay

Coenagrionidae: Ischnurinae
Enallagma novaehispaniae • Neotropical Bluet ♂
Metetí, Darién, Panama

Coenagrionidae: Ischnurinae
Oxyagrion rubidum • Coral ♂
Along Arroyo de las Víboras, Colonia, Uruguay

Coenagrionidae: Ischnurinae
Ischnura fluviatilis • Austral Forktail ♂
Colonia del Sacramento, Colonia, Uruguay

Coenagrionidae: Protoneurinae
Neoneura esthera • Esther's Threadtail ♂
Stream near Lagunas de Aruza, Darién, Panama

Coenagrionidae: Ischnurinae
Ischnura ramburii • Rambur's Forktail ♂
Oranjestad, Oranjestad, Aruba

Coenagrionidae: Protoneurinae
Neoneura confundens • Lemon-striped Threadtail ♂
Comarca Indígena Emberá-Wounaan, Panama

Coenagrionidae: Leptobasinae
Telebasis dominicana • West Indian Firetail ♀
Les Cayes, Sud, Haiti

Coenagrionidae: Pseudostigmatinae
Microstigma rotundatum • Helicopter Damsel ♂
Along Rio Negro, Amazonas state, Brazil

Coenagrionidae: Leptobasinae
Telebasis filiola • Striped Firetail ♂
Gamboa, Colón, Panama

Coenagrionidae: Pseudostigmatinae
Mecistogaster linearis • Long-tailed Helicopter ♀
Soberanía National Park, Colón, Panama

Coenagrionidae: Leptobasinae
Telebasis filiola • Striped Firetail ♀
Gamboa, Colón, Panama

Coenagrionidae: Pseudostigmatinae
Megaloprepus caerulatus • Blue-winged Helicopter ♀
Soberanía National Park, Colón, Panama

Lestidae: Lestinae

Archilestes grandis • Great Spreadwing ♂

Tuis, Cartago, Costa Rica

Platystictidae: Sinostictinae

Palaemnema sp. • Shadowdamsel ♂

Tuis, Cartago, Costa Rica

Lestidae: Lestinae

Lestes forficula • Rainpool Spreadwing ♀

La Ermita, Panamá Oeste, Panama

Heteragrionidae: Heteragrioninae

Heteragrion erythrogastrum • Red-and-black Flatwing ♂

Soberanía National Park, Colón, Panama

Platystictidae: Sinostictinae

Palaemnema sp. • Shadowdamsel ♂

Parque Nacional Pico Bonito, Atlántida, Honduras

Megapodagrionidae: Megapodagrioninae

Megapodagrion megalopus • Big-footed Flatwing ♂

Along Rio Negro, Amazonas, Brazil

Figure 7 **A view of the front of a great pondhawk, *Erythemis vesiculosa,* a species found throughout the Neotropics and as far north as the central United States. It has even been recorded in the Galápagos Islands.**

2

ORDER ORTHOPTERA

Grasshoppers, Crickets, and Katydids

The basic body plan of a grasshopper has been largely unchanged since the time of the dinosaurs. And just as odonates were the first creatures to take to the air, orthopterans were likely the first to produce song—in this case, for attracting the opposite sex.

The order Orthoptera is made up of two suborders. The Caelifera are the grasshoppers, and the Ensifera are the crickets and katydids.

Orthopterans' bodies are somewhat cylindrical. The middle pair of legs is large, and the hind pair very large, especially the femur, which is filled with muscles used for jumping. The compound eyes are large, and ocelli (simple eyes) are often present. The mouthparts are mandibulate for chewing. The antennae are filiform (thread-like) and are short in Caelifera and long in Ensifera.

Some groups of orthopterans are wingless as adults, or have very short wings. A majority of species, however, have two pairs of large wings. The forewings are thickened, and one important function of these is to protect the hindwings. The hindwings are membranous, and fold up to fit under the forewings when not in use. At rest, the wings are held over the back like a steep roof, or in some cases (especially among crickets) held flat over the back.

Food. Grasshoppers eat plant material. They may occasionally consume other things they happen to run across, but this group is phytophagous (plant feeding).

The Ensifera, on the other hand, are typically ominivores, and will eat living and dead plant material and dead arthropods and other small animals. A few groups among the katydids are predaceous, and have grasping front legs like a mantis. Underground species sometimes feed on roots, or consume mud that contains minute living things. Some mole crickets behave like rodents, collecting seeds and storing them underground for later consumption.

Reproduction. Many orthopterans use sound to find mates. The Ensifera, in particular, often have songs that sound musical to human ears. Generally it is males who sing, but in some species females may respond and the pair may in effect sing a duet. The two suborders produce souds differently. In Ensifera the bases of the forewings have a file and a scraper, and when the wing bases are rubbed together sound is produced. No Caelifera produce sound this way; instead, grasshoppers typically rub a peg on the hind leg against one or more prominent veins on the forewing.

Sound reception in the two suborders is also different. The sound-receiving organs of Ensifera are located on the forelegs, in the form of a visible hole that bears a membrane within it. The "ears" of grasshoppers, on the other hand, are at the base of the abdomen. Note that some orthopterans do not sing, while others produce song with a frequency too high for humans to hear.

In Orthoptera the usually-smaller male mounts the female, and the two may remain coupled for some time.

Development. Generally the female Ensifera have noticeable ovipositors, while the ovipositors of grasshoppers are smaller. A majority of grasshoppers lay their eggs in soil, while Ensifera lay their eggs in the soil, in a protective crevice, or on plants, depending on the group.

Orthoptera are hemimetabolous, having incomplete metamorphosis. The nymphs resemble the adults but lack wings. Because many orthopterans are wingless as adults, or short-winged, it can be difficult to distinguish nymphs from adults in some species, except by size. Orthopterans typically go through four to ten instars before reaching the adult stage. (An instar is the period between molts.)

Figure 8 This early instar katydid nymph has a tiny, wingless body but its back legs seem nearly adult-sized. Brazil.

Dangers to orthopterans. Orthopterans are generally robust and make protein-rich meals for many creatures, including occasionally humans. Birds and spiders are among the most important predators, but reptiles, amphibians, mammals, and other insects also eat their share. Fish will quickly make a meal of any grasshopper or cricket that enters the water.

Orthoptera also fall victim to entomopathogenic fungi, which can essentially take over and fatally cover the body with visible fungus. Parasitoids also attack some groups of grasshoppers, crickets, and katydids, but Orthoptera are not as vulnerable as some other groups like caterpillars.

(Parasitoids such as wasp larvae feed on their victims and eventually cause the death of their host.)

Figure 9 This tropical house gecko has been hunting indoors at night and captured some species of Ensifera. Dominican Republic.

Defenses. Though Orthoptera can walk, jump, fly, and many can swim well, they nearly all choose jumping as their first line of defense. These creatures can jump many times their body length. Flying is used more often for long distance dispersal than for defense. However, some orthopterans have elements on their hindwings (such as eyespots) that can be flashed to startle a predator. Some reveal bright colors on the hindwings as they fly, then suddenly abort their flight and drop to the ground. A predator who had fixated, for example, on red hindwings, suddenly is confused as to where that color has gone, and thus loses its prey.

Crypsis—blending in with a background, or mimicking something harmful—is relied on by some Orthoptera, especially katydids. Some katydids are remarkable mimics of green leaves, curled up dead leaves, or even the delicate branchings of lichen. Some early instar nymphs of Orthoptera mimic bird droppings, or ants. Predators would likely deem the former unworthy of notice, while judging the latter a source of danger.

A number of orthopteran groups are chemically protected and noxious to predators, and they warn predators by sporting aposematic coloration (often red or orange and black). For this reason one often sees Neotropical species that are brightly colored and that sit out in the open on leaves and come to no harm.

Relationship to humans. Orthoptera are an economically important group, with many species feeding on crops.

In the Neotropics the most devastating damage is done by species in genus *Schistocera*, which can behave as locusts and in huge numbers attack farmers' fields. At other times *Schistocera* species are solitary, though they still damage crops.

Crickets and katydids can be agricultural pests as well, but typically are less damaging than species in Caelifera.

Neotropical Specialties. Among the Caelifera, the Lubber Grasshopper subfamily Romaleinae frequently excites the interest of naturalists and scientists. These insects are doubly interesting because they are both large and colorful. Several genera of this subfamily are found in the southern half of the United States, but the Neotropic region is the best place to find these often huge grasshoppers. Their colors can include purple, orange, yellow, red, and green.

Various katydid groups also reach an impressive size in the Neotropics. One species popular among naturalists is the Rhinoceros Katydid, *Copiphora rhinoceros*, found in Central America. In addition to its large size, it is notable for having a prominent yellow horn on the head. Both males and females possess the large horn, which apparently is used to fend off attacks by bats. The genus *Copiphora* is found only in the Neotropics and in South Africa..

Ripipterygidae is a family found only in the Neotropics. These insects go by the name mud crickets. They are not crickets, however, but Caelifera. This odd group of grasshoppers are often blue-black with whitish stripes, sometimes with orange accents. They usually live beside streams, and have the remarkable abilty to jump from the surface of the water. They are quite small for Caelifera, with most species measuring 3 to 11 mm.

One strange-looking grasshopper group found only in South America and Panama is the Jumping Stick family, Proscopiidae. Like phasmids, these bear a close resemblance to sticks. Unlike phasmids, however, the jumping sticks have short antennae bearing few segments. The heads of jumping sticks are vertical, perpendicular to the stick-like body. Wings are absent or vestigial. Also missing in the jumping stick family are structures for producing sound or hearing it.

Figure 10 Comparison of a nymph of the collared lubber *(Tropidacris collaris)* with the adult. Structurally they are similar, though the nymph lacks wings and developed reproductive parts. As with many Orthoptera, the color and markings change quite a bit in the progression from nymph to adult. Brazil.

Another interesting Neotropical group are the monkey grasshoppers (Eumastacidae). These are found all over the American tropics and are easily distinguished from other groups because they hold their legs perpendicular to their bodies—in a bird's eye view they look like an airplane. In genera such as *Eumastax* the colors are often bright, and may include, for example, green, blue, and red on the same insect.

Figure 10a Stick grasshoppers (Proscopiidae) have heads perpendicular to their stick-like bodies. Note the half circle of feeding damage to the leaf. French Guiana.

Figure 10b Monkey grasshoppers with undeveloped wings can be tricky. Some are nymphs, but others are adults of wingless or short-winged species. Lake Machuwasi, Madre de Dios, Peru.

Acrididae: Leptysminae
Cornops sp. • Toothpick Grasshopper
Rewa village, Upper Takutu-Upper Essequibo, Guyana

Acrididae: Proctolabinae
Drymophilacris sp. • Nightshade Grasshopper nymph
Montagne de Kaw, Roura, French Guiana

Acrididae: Leptysminae
Tetrataenia surinama • Suriname Grasshopper
Novo Airão, Amazonas, Brazil

Acrididae: Proctolabinae
Lithoscirtus miniatulis • Proctolabine Grasshopper
Tuis, Cartago, Costa Rica

Acrididae: Leptysminae
Stenacris cf. fissicauda • Toothpick Grasshopper
Tortuguero, Limón, Costa Rica

Acrididae: Oedipodinae
Heliastus sumichrasti • Band-winged Grasshopper
La Ermita, Panamá Oeste, Panama

Acrididae: Oedipodinae
Trimerotropis sp. • Band-winged Grasshopper
Puerto Madryn, Chubut, Argentina

Acrididae: Ommatolampidinae
Anablysis teres • Short-horned Grasshopper
Berg en Dal, Brokopondo, Suriname

Acrididae: Ommatolampidinae
Abracris flavolineata • Yellow-lined Grasshopper
Parque Nacional Tayrona, Magdalena, Colombia

Acrididae: Ommatolampidinae
Eusitalces cf. vittatus • Short-horned Grasshopper
Along the Rupununi River, Guyana

Acrididae: Ommatolampidinae
Abracris sp. • Short-horned Grasshopper
Parque Natural Metropolitano, Panamá, Panama

Acrididae: Ommatolampidinae
Omalotettix obliquus • Oblique Omalotettix
Montsinéry, Montsinéry-Tonnegrande, French Guiana

Acrididae: Ommatolampidinae
Parasitalces sp. • Short-horned Grasshopper
Berg en Dal, Brokopondo, Suriname

Acrididae: Ommatolampidinae
Ommatolampis perspicillata • Conspicuous Grasshopper
Berg en Dal, Brokopondo, Suriname

Acrididae: Ommatolampidinae
Sciponacris sp. • Short-horned Grasshopper
Beside Lake Machuwasi, Madre de Dios, Peru

Acrididae: Ommatolampidinae
Syntomacris guttulosa • Spotted Syntomacris
Along the Rupununi River, Guyana

Acrididae: Ommatolampidinae
Xiphiola cyanoptera • Blue-winged Grasshopper
Réserve Naturelle Trésor, Roura, French Guiana

Acrididae: Ommatolampidinae
Leptomerinthoprora brevipennis • Grasshopper
Tuis, Cartago, Costa Rica

Acrididae: Rhytidochrotinae
Rhytidochrota sp. • Wrinkled-skin Grasshopper
Shipetiari indigenous community, Madre de Dios, Peru

Acrididae: Gomphocerinae
Orphulella concinnula • Slant-faced Grasshopper
Tortuguero National Park, Limón, Costa Rica

Acrididae: Gomphocerinae
Amblytropidia trinitatis • Slant-faced Grasshopper
Parque Natural Metropolitano, Panamá, Panama

Acrididae: Gomphocerinae
Orphulella punctata • Slant-faced Grasshopper
Along the Rupununi River, Guyana

Acrididae: Gomphocerinae
≈ Amblytropidia sp. • Slant-faced Grasshopper
Berg en Dal, Brokopondo, Suriname

Acrididae: Gomphocerinae
Orphulella sp. • Slant-faced Grasshopper
Reserva Particular do Cristalino, Mato Grosso, Brazil

Acrididae: Gomphocerinae
Compsacris pulcher • Beautiful Compsacris
Berg en Dal, Brokopondo, Suriname

Romaleidae: Romaleinae
Xyleus discoideus • Round-plated Lubber Grasshopper
Parque Nacional Tayrona, Magdalena, Colombia

Acrididae: Cyrtacantacridinae
Schistocerca flavofasciata • Bird Grasshopper
Along the Rupununi River, Guyana

Romaleidae: Romaleinae
Taeniopoda auricornis • Golden-horned Lubber
Cuero y Salado Refuge, Atlántida, Honduras

Romaleidae: Romaleinae
Prionolopha serrata • Serrated Lubber Grasshopper
Montsinéry, Montsinéry-Tonnegrande, French Guiana

Romaleidae: Romaleinae
Taeniopoda reticulata • Reticulated Lubber
La Suiza, Cartago, Costa Rica

Romaleidae: Romaleinae
Taeniopoda reticulata • Reticulated Lubber nymph
La Suiza, Cartago, Costa Rica

Romaleidae: Romaleinae
Agriacris magnifica • Magnificent Lubber
Santo Domingo de los Colorados, Pichincha, Ecuador

Romaleidae: Romaleinae
Tropidacris collaris • Collared Lubber
Novo Airão, Amazonas, Brazil

Romaleidae: Romaleinae
Agriacris tricristata • Three-crested Lubber
Sansón, Darién, Panama

Romaleidae: Romaleinae
Tropidacris cristata • Crested Lubber
Suchitoto, Cuscatlán, El Salvador

Romaleidae: Romaleinae
Colpolopha obsoleta • Obsolete Lubber nymph
Montsinéry, Montsinéry-Tonnegrande, French Guiana

Episactidae: Episactinae
Episactus cf. tristani • Episactid Grasshopper adult
Suchitoto, Cuscatlán, El Salvador

Eumastacidae: Eumastacinae
Eumastax cf. vittata • Striped Monkey Grasshopper
Along the Rio Negro, Amazonas, Brazil

Eumastacidae: Eumastacinae
Eumastax cf. pectipes • Striped Monkey Grasshopper
Manú National Park, Madre de Dios, Peru

Eumastacidae: Eumastacinae
Homeomastax sp. • Monkey Grasshopper nymph
Tuis, Cartago, Costa Rica

Proscopiidae: Prosopiinae
Proscopia sp. • Stick Grasshopper
Montagne de Kaw, Roura, French Guiana

Proscopiidae: Prosopiinae
Proscopia sp. • Stick Grasshopper
Along the Rio Negro, Amazonas, Brazil

Pyrgomorphidae: Pyrgomorphinae
Omura congrua • Gaudy Grasshopper
Berg en Dal, Brokopondo, Suriname

Ripipterygidae: Ripipteryginae
Ripipteryx forceps species group • Pygmy Mole Cricket
Nanegalito, Tandayapa Valley, Ecuador

Tetrigidae: Tetriginae
Stenodorus sp. • Pygmy Grasshopper
Berg en Dal, Brokopondo, Suriname

Gryllidae: Gryllinae
Gryllini • Field Cricket
Reserva Particular do Cristalino, Mato Grosso, Brazil

Ripipterygidae: Ripipteryginae
Ripipteryx limbata • Pygmy Mole Cricket
Tuis, Cartago, Costa Rica

Gryllidae: Podoscirtinae
Podoscirtinae • Bush Cricket
Berg en Dal, Brokopondo, Suriname

Gryllidae: Podoscirtinae
Tafalisca cf. lineatipes • Bush Cricket
Long Bay, Portland parish, Jamaica

Trigonidiidae: Trigonidiinae
Trigonidiini • Sword-tailed Cricket
Along the Rewa River, Guyana

Trigonidiidae: Nemobiinae
Nemobiinae • Ground Cricket
Along the Rupununi River, Guyana

Gryllidae: Eneopterinae
Eneoptera guyanensis • Brown Bush Cricket
Montsinéry, Montsinéry-Tonnegrande, French Guiana

Trigonidiidae: Trigonidiinae
Trigonidiinae • Sword-tailed Cricket
Tuis, Cartago, Costa Rica

Gryllidae: Eneopterinae
Eneoptera surinamensis • Suriname Bush Cricket
Along the Rio Negro, Amazonas, Brazil

Gryllidae: Eneopterinae
Eneopterinae • Bush cricket
El Hato, Sacatepéquez, Guatemala

Gryllotalpidae: Gryllotalpinae
Neocurtilla hexadactyla • Northern Mole Cricket
Reserva Particular do Cristalino, Mato Grosso, Brazil

Phalangopsidae: Phalangopsinae
Aclodes cf. nicuesa • Phalangopsid Cricket
Tuis, Cartago, Costa Rica

Tettigoniidae: Phaneropterinae
Ceraia sp. • Phaneropterine Katydid
Reserva Particular do Cristalino, Mato Grosso, Brazil

Tettigoniidae: Phaneropterinae
Euceraia femorata • Phaneropterine Katydid
Berg en Dal, Brokopondo, Suriname

Tettigoniidae: Phaneropterinae
Microcentrum sp. • Angle-winged Katydid
Leona Vicario, Quintana Roo, Mexico

Tettigoniidae: Phaneropterinae
≈ Scudderia sp. • Scudder's Katydid nymph
Along the Rupununi River, Guyana

Tettigoniidae: Phaneropterinae
Philophyllia sp. • Phaneropterine Katydid
Parque Nacional Tayrona, Magdalena, Colombia

Tettigoniidae: Phaneropterinae
≈ Scudderia sp. • Scudder's Katydid nymph
Parque Nacional Pico Bonito, Honduras

Tettigoniidae: Phaneropterinae
≈ Phylloptera sp. • Phaneropterine Katydid
Manú National Park, Madre de Dios, Peru

Tettigoniidae: Phaneropterinae
Orophus tessellatus group • Leaf Katydid
Comarca Indígena Emberá-Wounaan, Panama

Tettigoniidae: Phaneropterinae
Phylloptera cf. festae • Festive Leaf Katydid
Along the Suriname River, Brokopondo, Suriname

Tettigoniidae: Phaneropterinae
Stilpnochlora thoracica • Leaf Katydid
Mountain Pine Ridge Forest Reserve, Belize

Tettigoniidae: Pseudophyllinae
Melanonotus powellorum • Short-winged Katydid
Monteverde, Puntarenas, Costa Rica

Tettigoniidae: Phaneropterinae
Stilpnochlora sp. • Leaf Katydid
Zitácuaro, Michoacán, Mexico

Tettigoniidae: Pseudophyllinae
Leptotettix cf. spinoselaminatus • Katydid
Along the Rewa River, Guyana

Tettigoniidae: Phaneropterinae
Pycnopalpa bicordata • Leaf Katydid
Tuis, Cartago, Costa Rica

Tettigoniidae: Conocephalinae
Conocephalus sp. • Meadow Katydid
La Ermita, Panamá Oeste, Panama

Tettigoniidae: Conocephalinae
Copiphora rhinoceros • Rhinoceros Katydid
Monteverde, Puntarenas, Costa Rica

Tettigoniidae: Conocephalinae
Neoconocephalus sp. • Common Conehead
Sansón, Darién, Panama

Tettigoniidae: Conocephalinae
Vestria sp. • Conehead Katydid
Berg en Dal, Brokopondo, Suriname

Tettigoniidae: Conocephalinae
Neoconocephalus sp. • Common Conehead
Long Bay, Portland parish, Jamaica

3

ORDER PHASMATODEA

Stick Insects

Entomologists who specialize in stick insects must sometimes curse their choice. These insects can be extremely hard to find! Many bear an almost perfect resemblance to very small branchlets, and the insect will sway its body to match the swaying of the plant. Some groups resemble leaves instead—brown or green—and these insects, too, can blend in nearly undetectably. Collecting days can go very slowly for Phasmatodea researchers.

The order is a relatively small one, with some 3000 mostly tropical species described thus far worldwide. Of these, some 800 are Neotropical. Stick insects have mandibles for chewing, and the head is directed generally forward. The compound eyes are relatively small, and most species lack ocelli (simple eyes). Antennae are of various lengths, with from 8 to 100 antennomeres (segments). Unlike mantises, stick insects have a short prothorax.

Many species are wingless, while in others the females have reduced wings and only the males can fly. In winged species the forewings are leathery and reduced in size, while the hindwings are membranous but with the foreward edge often thickened. The wings are folded flat over the back when not in use. Many walkingsticks are giants of the insect world in terms of length. The longest insect in the world is an Australian stick insect that tops 56 cm (22 inches) in length, including the habitually stretched out forelegs. The longest insect in the Americas is a Neotropical phasmid, *Heteronemia grande*, which has a body length of up to 26 cm (10.2 inches). By weight, though, no stick insect is even in the running for largest insect.

This interesting order goes by a variety of names, both scientific and common. Specialists can't agree whether the best name for the order is Phasmatodea or Phasmida. The rigorous rules for naming species, genera, and families are relaxed when it comes to orders and taxa above, so both names are valid. Similarly, when it comes to English common names, some people prefer walkingstick, while many others, including nearly all British naturalists, prefer stick insect.

Food. Stick insects are strict herbivores. Often they seek out only one species of plant, or several related species, to provide their food as well as the ideal place to deploy their particular camouflage. These insects feed at night. They rest—sometimes out in the open but camouflaged—in the daytime.

Reproduction. Since the females of most species do not fly, and since the nearest male may be a long distance away, female stick insects of many species use pheromones to attract mates. Males that detect the pheromone can then fly in to the female's calling location. With few if any courtship niceties, the smaller male mounts the female and bends his abdomen 180 degrees so that its tip can meet the female's abdomen. Most males have some sort of gripping mechanism to insure the connection is not lost--not lost for hours or even days. The long period of attachment is the male's way of ensuring no other male's sperm will be used to fertilize any of the female's eggs. In the species with long coupling, the female soon resumes her normal activities, and the little male simply rides along.

In most stick insect species the female drops the eggs one by one onto the forest floor or other substrate. The eggs look like small, hard seeds, and are probably not recognized by wasps or flies looking for eggs to parasitize. Females that use this broadcast method typically produce a lot of eggs—up to 1300. In other groups, the female attaches the eggs to a favored food plant. Some *Anisomorpha spp.* dig a hole and bury their eggs as added insurance against predation. And some species produce eggs that are capped with a tasty treat, leading ants to carry the eggs to their nest where the eggs are sheltered from predators and parasitoids. The ants eat the tasty part of the egg, and lay the rest aside. The stick insects leave the ant colony after hatching.

In cases where a female cannot find a male, some walkingstick species can reproduce parthenogenetically. This usually results in an all-female batch of nymphs, but in some species parthenogentic reproduction can lead to hatchlings of both sexes.

Development. Walkingstick eggs often have a "lid" which the hatchling pushes open to emerge. The young nymphs have evolved to move toward bright light, and thus their first action is often climbing a tree or herbaceous plant. Walkingsticks develop by incomplete metamorphosis, in which the immatures look very much like the adults, though smaller and always wingless. In many species in this order, males go through five instars, and females six. Thus males become adults sooner, but they also require about two weeks more to reach sexual maturity. The females are sexually mature immediately after the final molt, and thus both sexes reach sexual maturity at about the same time.

Figure 11 A mating pair of *Malacomorpha jamaicana* at Long Bay, Jamaica. The pair will remain coupled for some time, with the larger female going where she chooses and the smaller male going along for the ride.

Figure 12 A recently molted early instar nymph of *Anthericonia anketeschkei* (Pseudophasmatidae). Such insects are in danger of predation because of their soft integument.

Dangers to Stick Insects. Despite their excellent camouflage, stick insects are attacked by predators ranging from birds to ants. Certain wingless parasitoid wasps attack the eggs of some phasmids. Biting midges in the genus *Forcipomyia* sometimes attack stick insects and pierce their wings or integument to feed on the haemolymph.

Defenses. The first line of defense against predators is what entomologists call "masquerade crypsis"—masquerading as twigs, branchlets, or leaves. As a backup plan, a few species have bright colors or patterns on the hindwings that can be flashed suddenly to startle an attacker.

If these relatively benign defenses don't work, stick insects turn to chemical warfare. Chemical production glands, unusually located on the prothorax, manufacture a noxious compound that is sickening to any creature receiving it. Some stick insects can actually spray the chemical, aiming it for the greatest effect. At least one mammal group has learned to cope with the chemical assault. Some mouse possums (a Neotropical group, *Marmosa spp.*) have learned to hold the stick insect in such a way that the spray is aimed away from them. When the spray is exhausted, the possum begins its meal.

Walking sticks are not what you would call born fighters, but they can defend themselves with their mandibles and their sometimes spiny legs. Like mantises, they are quite willing to drop a leg in a fight (the leg detaches along a built-in weak spot). In the case of nymphs, the leg can reappear at the next molt.

Like many other insects, stick insects employ the defense of dropping to the ground, where the predator must try to locate them again. Playing dead also leads some predators to lose interest.

Relationship to humans. Phasmatodea have only the most tenuous relationship with humans. The few people who see one in the wild are likely to pick it up and show it to others. Entomologists and hobbyists put them in collections, or keep them as pets or in the laboratory. Economically important Phasmatodea are found only in

Australia, where two species are noted for defoliating eucalyptus plantations.

Neotropical Specialties. The New World tropics, and Old World tropics of the southern hemisphere, are the best places to see a diversity of stick insects. Temperate areas such as Europe, Siberia, Canada, and Chile have a paucity of species.

One trait of stick insects that seldom fails to delight naturalists is length. A number of Neotropical species are impressively long, surpassing 20 cm (8 inches) in length. One fun species to observe is *Bostra lobata*, which is not only long but high as well, with six long, gangly legs. This species sports two horns on the head. Another intesting species in the same family (Diapheromeridae) is *Oncotophasma maculosum*, which sports green rings on its brown body, particularly where the legs join. Colors can change with the perch location.

A truly bizarre member of the Neotropic fauna is *Prisopus berosus* in the family Pseudophasmatidae. On first sight it is hard to tell not only which family, but which order this insect belongs to. It is somewhat soft-bodied, and its colors are muted, not glossy. It is not thin or cylindrical like so many other stick insects. The soft green and brown colors are attractive, but what makes this insect so strange-looking are the various flanges, warts, hairs, and saw-toothed edges. This is masquerade crypsis to blend in with—what? Leaves, twigs, sticks, vines, grasses, or all of the above? The species is found from Oaxaca to the Yucatán, and south to Panama.

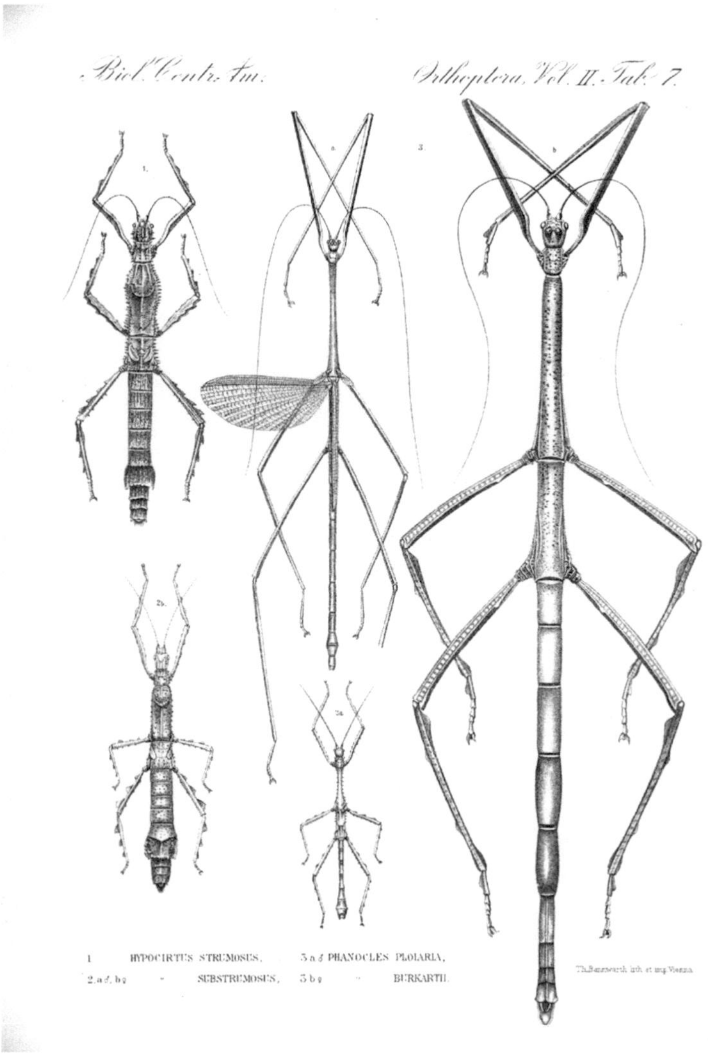

Figure 13 The volume of *Biologia Centrali-Americana* that covers stick insects can still be useful in identifying Mexican and Central American species. Unfortunately, by this late in the work's publication history (1900-1909), most of the work's plates were in black and white.

Pseudophasmatidae: Pseudophasmatinae
Pseudophasma flavipes ● Yellow-footed Pseudophasma
Montsinéry, Montsinéry-Tonnegrande, French Guiana

Pseudophasmatidae: Pseudophasmatinae
Pseudophasma unicolor ● Unicolor Pseudophasma
Altos del Maria, Coclé, Panama

Pseudophasmatidae: Pseudophasmatinae
Malacomorpha jamaicana • Jamaica Walkingstick pair
Long Bay, Portland parish, Jamaica

Pseudophasmatidae: Pseudophasmatinae
Prisopus berosus • Saw-footed Walkingstick
Tuis, Cartago, Costa Rica

Pseudophasmatidae: Stratocleinae
Parastratocles sp. • Striped Walkingstick
Altos del Maria, Coclé, Panama

Pseudophasmatidae: Xerosomatinae
Xylospinodes jensjohannseni • Johannsen's Pseudophasmatid
Monteverde, Puntarenas, Costa Rica

Diapheromeridae: Diapheromerinae
Bostra lobata • Lobed Walkingstick
Tuis, Cartago province, Costa Rica

Diapheromeridae: Diapheromerinae
Oncotophasma cf. maculosum • Maculate Walkingstick
Monteverde, Puntarenas, Costa Rica

Diapheromeridae: Diapheromerinae
Oncotophasma martini • Martin's Stick Insect
Parque Natural Metropolitano, Panamá province, Panama

4

ORDER MANTODEA

Mantises

Mantises are the owls of the insect world. Sitting still with large eyes peering, they give the impression—deserved or not—of great wisdom. The name mantis is derived from the Greek word for soothsayer. Like an owl, the mantis holds its body completely still while swiveling its head in a wide arc. After a great deal of waiting, both mantis and owl move suddenly and silently to attack their prey. Both creatures eschew other kinds of nourishment, and feed only on fresh-caught food. Both are excellent examples of a sit-and-wait predator.

Figure 14 Portrait of a hooded mantis by Maria Sibylla Merian. Merian was German by birth and Dutch by residence. In 1699 she and her daughter traveled to Suriname, then a Dutch colony, so she could study and paint insects and plants there. She was one of a small number of scholars of that time who studied and wrote about metamorphosis.

Mantises are closely related to insects in Blattodea, the cockroaches and termites. Like those creatures, mantises have mandibulate (chewing) mouthparts that are directed downwards. The head, though, is triangular in mantises, and has very large eyes placed high up on the head. A dark spot on each eye often resembles a pupil. Nearly all male mantises are fully winged but in a number of species the females are short-winged or wingless. Unlike cockroaches or stick insects, mantises have an unusually long thorax (the prothorax in particular) which gets their soft abdomen away from the attacks being conducted by the head.

Mantodea is not a particularly speciose order. There are about 2500 described species worldwide with about 500 of those found in the Neotropics. Insects in this order are called mantises in English, or, if they are in the large family Mantidae, are sometimes called mantids. The name praying mantis is also used for various species, based on the apparently religious posture of most mantises; "preying mantis" would be equally apt.

Food. Mantises are insectivorous. They eat what they kill, and their prey is almost exclusively insects—and occasionally other arthropods such as spiders. Internet photos document that mantises are capable of capturing hummingbirds, small fish, frogs, and lizards. One researcher has reported mantis capture of a deer mouse.

Figure 15 A hooded mantis *(Choeradodis rhomboidea)* devours its moth prey. Panama.

Locating food is facilitated by a number of Mantodea's anatomical features. One is the widely spaced front-facing eyes, in an ideal position for three-dimensional vision that facilitates motion detection. The forelegs are weapons designed to immobilize prey as the spines impale. Pierced by the spines of two forelegs and gripped by the mandibles, escape from a mantis attack is unlikely. Humans, too, by the way, can be hurt by the mantis's sharp spines.

Like cats, mantises are excellent at detecting the slightest motion. On the other hand, researchers have reported that mantises sometimes fail to see insect prey that is in plain sight because that insect is standing perfectly still.

Reproduction. The oft-repeated factoid about mantis reproduction is that during mating the female may turn her head toward the rear and begin eating her mate's head. The "punchline" of this story, which people seem to enjoy telling, is that despite the loss of his brain, the male continues to mate for up to an hour.

Some researchers, however, believe that such attacks have been documented largely in captive specimens, and may be related to factors like underfeeding or overcrowding. It is not clear, at any rate, how often this decapitation happens in mating in the wild, or in how many mantis species it may occur. It may be that one reason mantis males are often considerably smaller than the female is that it means the female cannot easily reach the male while the two are joined at the tips of their abdomens.

The female often is the one that advertises for a mate, either by pheromones or by stridulation. It may be that it is the female that calls because females are wingless in many species, and cannot fly and look for a mate.

In many cases a single mating suffices for the lifetime of a female. She stores the sperm, which will remain viable for four or five months and can be used in multiple instances of egg-laying. Mantises lay their eggs in double rows. The eggs are enclosed in a protective froth that hardens quickly. Sometimes the frothy ootheca is attached to a twig or plant stem while in other instances it is attached to a vertical surface such as a tree trunk or wall.

Development. The ootheca contains from 10 to 400 eggs that typically hatch in about a month. In temperate areas the egg case may overwinter before the eggs hatch. The hatchlings exit via one-way valves in the ootheca, entering the most vulnerable period of their lives. At this stage neither their mouthparts nor their leg spines are large enough to be deadly to any but the smallest insects. Many early instar nymphs resemble ants, a type of insect that few other arthropods care to attack. As they molt and get larger, the mantis nymphs may change appearance considerably, since the ant masquerade will no longer work at larger sizes.

Figure 16 The female mantis lays eggs that are surrounded by a froth that soon hardens and provides a tough, weather-resistant shelter for the eggs. Brazil.

Most mantis species go through six or seven instars.

Dangers to mantises. As for predation on mantises, it is safe to say that far more insects die from adult mantis attacks than adult mantises die from the attacks of other insects. Nevertheless, like all insects, mantises are preyed upon such creatues as birds, reptiles, amphibians, mammals, and occasionally other arthropods. Nymphs are at more risk of predation than adults.

Adult mantises are not particularly known for widespread dieases or attacks by parasites or parasitoids. Mantis oothecas, though, may be attacked by parasitoid wasps such as those in the genus *Podagrion*.

Defenses. As with other kinds of sit-and-wait predators, mantises rely on camoflage as a primary defense. Their green and brown coloration is useful for hiding amid living or dead plants, and many species can switch from green to brown (or vice versa) fairly quickly. A smaller number of species blend in with colorful flowers, and some have wings that resemble a large green or brown leaf, right down to the fake mildew and insect holes.

Figure 17 A female *Podagrion sp.* wasp (Torymidae) oviposits in a mantis ootheca. Nearby are exits holes, possibly of the same wasp species.

Nymphs often blend in well with plant parts, and as noted in some species the nymphs mimic ants to scare off predators. In a small number of species maternal guarding helps defend eggs or hatchlings.

Mantises also defend themselves by counter-attacking with their forelegs and mouthparts. They are quite willing to lose one of the four rear legs in battle, since better to lose a limb than a life. These legs have a built-in weak point where they detach fairly easily. Nymphs will see these legs reappear at the next molt, but adult mantises cannot regenerate limbs.

Mantises are reasonably good fliers, and taking flight can result in their escape from ground-dwelling predators.

Figure 18 This nymph of an *Acontista sp.* bears a strong resemblance to an ant, with the result that many predators will leave the nymph alone. Manaus, Brazil.

Relationship to humans. Typically, mantises have little relation to humans, and that means they aren't usually subject to direct human attack via insecticides or introduced predators. Their lack of economic harm or benefit means that mantises have not been as well-studied as grasshoppers or beetles. Many humans find mantises interesting and photograph them, collect them, or keep them as pets.

Mantis eggs cases are marketed to gardeners, but purchasing them does not seem to result in much benefit. For one thing, the mantises may leave the garden and prefer to hunt in the dense weeds. For another, while it is true mantises will eat pest species, they also eat beneficial species like honey bees and other pollinators.

Neotropical Specialties. When someone from a temperate area visits the New World tropics, one thing quickly noticed is the increased diversity in many creatures including mantises. Insect watchers in France can observe eight species of mantis, but if they travel to the "overseas department" of French Guiana (a much smaller area) they can study ninety-three species. Sacré bleu!

One group of small mantises absent from Europe and Canada and all but the far south of the United States are the bark mantises in the genus *Liturgusa.* These are sometimes colored to blend in with tree bark, sometimes colored to match the lichens and mosses that grow on tree bark. Bark mantises station themselves on tree trunks, head down, and wait for prey to show up. If a predator approaches their favored spot they move to the other side of the tree like a squirrel.

Another genus with interesting appearance and behaviors is *Mantoida*, the only genus in family Mantoididae. These mantises are the exception to sit-and-wait predation. Unlike others in Mantodea, they actively run down their prey. Since they use a different style of hunting, their forelegs are atypical for the order—not as massive, and bearing fewer and shorter spines. This family is found only in the Neotropics (and parts of Florida), and in fact is is missing from the temperate areas of southern South America.

Leaf Mantises (Acanthopidae) are found between roughly the Tropic of Cancer and the Tropic of Capricorn. These remarkable insects look precisely like a large leaf—green for those roosting in trees, or brown for those hiding among leaf litter. Some of the most remarkable leaf mimics are in the genera *Acanthops* and *Metilia*. This family also has some of the best ant-mimics among early instar mantises.

One final Neotropical group worthy of mention are mantises in subfamily Vatinae in family Mantidae, which are found only in the Americas, from the far south of the U.S. to northern Argentina. These insects often have small flanges or extensions on their legs that help them blend in with lichens, mosses, tiny vines, and the like. Some bear a prominent horn on the head.

Figure 18a With their large, widely separated eyes, mantises have excellent 3D vision and can detect the slightest motion of prey species. This is a bark mantis *(Liturgusa sp.)* photographed in Mato Grosso, Brazil.

Mantoididae: Mantoidinae
Mantoida sp. • Mantoidid Mantis
Montagne de Kaw, Roura, French Guiana

Acanthopidae: Acanthopinae
Metilia brunneri • Dead Leaf Mantis
Montagne de Kaw, Roura, French Guiana

Acanthopidae: Acanthopinae
Metilia sp. • Leaf Mantis
Reserva Particular do Cristalino, Mato Grosso, Brazil

Acanthopidae: Stenophyllinae
Acontista sp. • Acanthopid Mantis late instar nymph
Novo Airão, Amazonas, Brazil

Liturgusidae: Liturgusinae
Liturgusa sp. • Bark Mantis
Berg en Dal, Brokopondo district, Suriname

Liturgusidae: Liturgusinae
Liturgusa sp. ● Bark Mantis nymph
Colakreek, Para, Suriname

Liturgusidae: Liturgusinae
Liturgusa sp. ● Bark Mantis nymph
Mountain Pine Ridge Forest Reserve, Belize

Liturgusidae: Liturgusinae
Liturgusa sp. ● Bark Mantis
Tuis, Cartago, Costa Rica

Mantidae: Vatinae
Vates sp. ● Vatine Mantis
Gamboa, Colón, Panama

Mantidae: Vatinae
Zoolea sp. • Horned Mantis
Montsinéry, Montsinéry-Tonnegrande, French Guiana

Mantidae: Stagmatopterinae
Oxyopsis sp. • Stagmomantine Mantis
Berg en Dal, Brokopondo district, Suriname

Mantidae: Stagmomantinae
Stagmomantis sp. • Stagmomantine Mantis
Crooked Tree Preserve, Belize district, Belize

Mantidae: Stagmomantinae
Stagmomantis cf. heterogamia • Stagmomantine Mantis nymph
Mountain Pine Ridge Forest Reserve, Belize

Mantidae: Stagmatopterinae
Parastagmatoptera cf. flavoguttata • Stagmomantine Mantis
Berg en Dal, Brokopondo district, Suriname

Mantidae: Choeradodinae
Choeradodis rhomboidea • Rhomboid Hooded Mantis
Gamboa, Colón, Panama

Mantidae: Choeradodinae
Choeradodis cf. rhomboidea • Rhomboid Hooded Mantis nymph
Reserva Particular de Cristalino, Mato Grosso, Brazil•

Thespidae: Thespinae
Macromusonia sp. • Thespid Mantis nymph
Reserva Particular do Cristalino, Mato Grosso, Brazil

Thespidae: Oligonicinae
Thrinaconyx fumosa ● Smoky-winged Thespid
Tuis, Cartago, Costa Rica

Thespidae: Thespinae
Musonia sp. ● Thespid Mantis
Playa Los Angeles, Magdalena department, Colombia

Thespidae: Pseudomiopteryginae
≈ Pseudomiopteryx guyanensis ● Thespid Mantis
Along the Rupununi River, Upper Takutu-Upper Essequibo, Guyana

5

ORDER BLATTODEA

Cockroaches and Termites

The half dozen species of cockroach that invade our kitchens have given insects in this order a bad name. Most people think of roaches as filthy brown pests active at night while honest folk are in bed. Yet nearly all species live inoffensive lives under leaf litter, beside ponds, or on desert sands. Some live below ground in caves, while others live high in the canopy of tropical forests. Despite their reputation as filthy, no one has yet demonstrated that cockroaches transmit diseases to humans.

Until fairly recently, termites were placed in their own order, but they are in fact very close structurally to certain cockroaches, particularly in those in the family Cryptocercidae (the brown-hooded cockroaches). The most important differences between termites and other Blattodea are related to life histories rather than anatomy.

Features shared by insects in this order include mandibulate (chewing) mouthparts similar to those of grasshoppers. The pronotum forms a large protective shield covering much of the thorax and often the head as well. Antennae are filiform (threadlike) and have many segments. The forewings are thicked into tegmina and lie flat over the insect's back, protecting the membranous hindwings. Wings are short or absent in some adult Blattodea.

Food. The primary food of termites, and of many cockroaches, is wood. Various cockroaches also feed on many other kinds of food, including dung, fungi, carrion, and rotting leaves. No insect in Blattoda specializes in predation, but a number of species will occasionally attack and consume other insects, including nymphs of their own species.

Because cellulose is difficult for any animal to digest, termites and some cockroaches rely on various types of microorganisms in their gut to produce enzymes that help break down cellulose and other components of a wood-based diet.

Reproduction in cockroaches. There is a wide array of reproductive practices among cockroaches. Members of some species call for prospective mates by stridulating (rubbing body parts together to make sound). Others call with pheromones. In some cases, males raise their wings and reveal glands that produce tasty food for females to nibble on. In 1999 researchers O. Zompro and I. Fritzsche revealed to the world the use of "headlights" (two luminous pronotal spots) by males of a Brazilian species of blaberid cockroach seeking females.

Figure 19 Though she painted more butterflies than other insects, Maria Sibylla Merian did find the time to study and paint these cockroaches *(Periplaneta sp.)*. She has included both an adult and a nymph. Suriname, circa 1700.

Despite all the effort to find mates, many cockroach females mate only once in their lives. The sperm is stored and can be used to fertilize subsequent batches of eggs, for those females who produce more than one batch. Many males, too, mate seldom or only once or indeed not at all. And in a few species of cockroach females can reproduce asexually through parthenogenesis.

Cockroaches mate in a tail to tail position. The female soon produces an ootheca—a leathery case that holds the eggs in two parallel rows inside. She may hide the ootheca, or carry it partly protruding from the apex of her abdomen, or carry it internally until hatching, at which time the nymphs emerge from her body in a type of live birth.

Development in cockroaches. The mothers sometimes stay with the hatchlings to help protect them. In two taxa both parents stay with the nymphs for much or all of their develoment. In some cases mothers actually feed the nymphs with glandular secretions. In many other instances, though, the nymphs are essentially on their own to find food and protect themselves from harm. Nymphs pass through five to fourteen instars depending on the species.

Figure 20 This Ectobiid cockroach is having a difficult molt. Normally the old exoskeleton would have split down the center of the dorsum, but that has not happened. The cockroach is managing to exit through the rear of its larval "skin."

Termites. People often have the misconception that ants and termites are closely related and that it is hard to tell them apart. Ants and termites are in widely separated orders and differ sharply as to life histories and anatomy. For example, ants are characterized by complete metamorphosis, while termites' metamorphosis is incomplete, with no pupal stage. As for anatomy, in ants the body is clearly divided into three sections, while in termites it looks like there are only two large body sections. Ants have elbowed antennae while termites do not.

Termites are eusocial. That is, a single female (or a small caste of reproductive females) produces all the offspring, while non-reproductives care for the young and do the other work of the colony. Termites always live in colonies inside structures they have engineered—towers, for example, or elaborate subterranean burrows. At the center of the colony are a king and queen, which mate repeatedly during their lifetimes. They produce male and female offspring, which usually remain nymphs for their entire lives. Unlike the helpless larvae found in ant colonies, the immature termites fill roles as workers or soldiers.

Across tropical America, one common type of termite nest looks like a dark brown basketball and is found in trees and on masonry walls. These are built by nailhead termites in genus *Nasutitermes*. Outside the nest are covered walkways that protect the workers as they head out to search of food or as they return. Soldier and worker termites are largely blind, and make good use of their antennae to find their way in the world.

Figure 21 A spherical nest of nailhead termites *(Nasutitermes sp.)* from St. John in the U.S. Virgin Islands.

An interesting question is, how and why did termites diverge from other Blattodea? Actually, one family of cockroach has a life history similar in many ways to that of termites, and according to recent taxonomic studies these two groups descended from a common cockroach ancestor. This family is Cryptocercidae, and it contains only one genus, *Cryptocercus*. Cockroaches in this group are classified as subsocial as opposed to eusocial.

Cryptocercus spp. construct elaborate nests, with tunnels, chambers, galleries, and partitions. Their building material is a mixture of their fecal matter and masticated wood. They spend a lot of time keeping their nests free of fungal growths, dead termites, and fecal pellets. Nymphs in this genus of cockroach often stay juveniles for a long time (up to 13 instars) and the development can be further delayed by low-protein diets, or by injuries such as those caused by a nestmate nibbling off the antennae. Thus the building of elaborate homes, the construction material, the incessant housekeeping of the nest, and the greatly prolonged juvenile period, are all traits of cryptocercid roaches that are shared by termites. This supports the idea that termites and cryptocercids descended from a common ancestor.

Another likely reason termites evolved a eusocial society is that the nymphs lose the beneficial microorganisms in their gut at each molt. So the termites have specialized so that the king and queen handle reproduction itself, and they can produce large numbers of offspring because they are freed from caring for the young. The nymphs, meanwhile, care for other nymphs, and most importantly they replenish the microbiome of the other nymphs' digestive systems. After a molt, a termite nymph is fed by another nymph substances that contain the needed microbes.

Dangers to Blattodea. Humans pose a great danger to both cockroaches and termites, given that cockroaches are considered loathsome kitchen pests and termites can do great damage to homes and outbuildings. Order Blattodea helps keep pest control companies in business.

Figure 22 This nest of *Amitermes sp.* in Amazonas, Brazil, is beginning to flood, but the termites will cope. Its design evolved to shed rainwater.

And of course, all Blattodea face the usual dangers of predation by birds, reptiles, amphibians, and mammals. Four species of anteaters (all Neotropical) are especially fond of termites. The greatest enemy of termites is not anteaters, though, but ants—a case of the eusocial attacking the eusocial. Cockroaches, too, are regularly attacked by army ants and other predatory species.

Defenses. The defenses deployed by termites and cockroaches are numerous and varied. Some species are protected by chemical defenses. Moreover, some species that live in close proximity to their conspecifics can release an alarm pheromone that causes all to move quickly to the nearest crevice—their flattened body type allows them access, but denies entrance to a bulkier predator. Simple camouflage is also a defense, as for example brown cockroaches are hard to see in leaf litter. Rounding out the list of defenses is the tactic of playing dead—many predators are uninterested in dead creatures. At least one species of cockroach backs up its acting skills by releasing a chemical that has the smell of a dead creature.

Termites in particular have an entire caste dedicated to defense. These soldier termites have large, sharp mandibles, or, depending on the species, one or more kinds of chemical weapon. And of course the termite towers, burrows, or spherical nests are useful in providing a relatively safe place.

Figure 23 The covered walkways outside *Nasutitermes sp.* nests protect the termites from storms, desiccation, and predators. Brazil.

Neotropical Specialties. Interesting for their defensive capabilities are nailhead termites in subfamily Nasutitermitinae, which are absent from Europe, northern Asia, Canada, and 48 of the U.S. states. The soldiers of this group have a spherical head that bears a prominent nozzle used to shoot glue onto an attacker, usually stopping the enemy in its tracks.

Naturalists are often pleasantly surprised at the colorful, day-active species in cockroach subfamily Pseudophyllodromiinae. Several of these are shown in the following pages. But undoubtedly the most incredulity shown by humans are reserved for the giant roaches in genera *Blaberus* and *Epilampra.* The two taxa are widespread in Mexico, the Caribbean, and Central and South America. Roaches in *Blaberus* can measure 89 mm in length (3.5 inches). They can be especially startling when flying just above the heads of a group of people. They are not a pest species, but may enter houses because of their attrraction to electric lights.

Blattidae: Blattinae
Periplaneta americana • American Cockroach
Long Bay, Portland parish, Jamaica

Ectobiidae: Pseudophyllodromiinae
Euphyllodromia sp. • Redheaded Cockroach
Réserve Naturelle Trésor, Roura, French Guiana

Blattidae: Blattinae
Pelmatosilpha sp. • Cockroach
Tuis, Cartago, Costa Rica

Ectobiidae: Pseudophyllodromiinae
Euphyllodromia cf. propinqua • Redheaded Cockroach
Along the Rio Negro, Amazonas, Brazil

Ectobiidae: Pseudophyllodromiinae
Euphyllodromia sp. • Redheaded Cockroach
Tuis, Cartago, Costa Rica

Ectobiidae: Pseudophyllodromiinae
Chorisoneura sp. • White-veined Cockroach
Montagne de Kaw, Roura, French Guiana

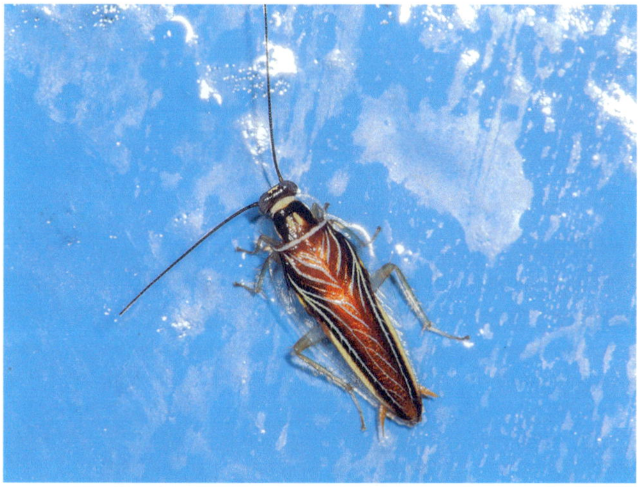

Ectobiidae: Pseudophyllodromiinae
Chorisoneura sp. • White-veined Cockroach
Berg en Dal, Brokopondo, Suriname

Blaberidae: Epilamprinae
Epilampra nr. sodalis • Ovoviviparous Cockroach
Novo Airão, Amazonas state, Brazil.

Ectobiidae: Pseudophyllodromiinae
Chorisoneura sp. • White-veined Cockroach
Long Bay, Portland Parish, Jamaica

Blaberidae: Epilamprinae
Epilampra sp. • Ovoviviparous Cockroach
Reserva Particular do Cristalino, Mato Grosso, Brazi

Ectobiidae: Blattellinae
Pseudomops sp. • Field Cockroach
Along the Rupununi River, Guyana

Blaberidae: Epilamprinae
Epilampra sp. • Ovoviviparous Cockroach
Sansón, Darién, Panama

ridae: Blaberinae

Blaberus sp. • Ovoviviparous Cockroach

Arima Valley, Trinidad, Trinidad and Tobago

Termitidae: Syntermitinae

Syntermes sp. • Mound-building Termites

Reserva Particular do Cristalino, Mato Grosso, Brazil

Blaberidae: Blaberinae

Blaberus sp. • Ovoviviparous Cockroach

Salvador, Manú National Park, Peru

Termiitidae: Nasutitermitinae

Nasutitermitinae • Nailhead Termite soldier

Montsinéry, Montsinéry-Tonnegrande, French Guiana

Blaberidae: Panchlorinae

Panchlora nivea • Banana Cockroach

Montagne de Kaw, Roura, French Guiana

Termitidae: Nasutitermitinae

Nasutitermitinae • Nailhead Termite workers and soldiers

Reserva Particular do Cristalino, Mato Grosso, Brazil

Ectobiidae
Ectobiidae ● Ectobiid Cockroach nymph
Berg en Dal, Brokopondo, Suriname

Hemiptera
Fulgoridae: Lystrinae
Lystra lanata • Fulgorid Planthopper
Montsinéry, Montsinéry-Tonnegrande, French Guiana

6

ORDER HEMIPTERA

TRUE BUGS, CICADAS, HOPPERS, AND APHIDS

Order Hemiptera includes insects that seem to have little in common. Stink bugs, water striders, cicadas, and aphids are all hemipterans. What traits unite these varied creatures?

The most important features that hemipterans have in common are double-tubed mouthparts used for piercing and sucking. The basic body plan of all insects includes two kinds of paired mouthparts, the maxillae and the mandibles. In Hemiptera the four structures are joined to form a cylindrical configuration containing two piercing blades and two tubes. The blades are used to cut into the food, of whatever kind it may be. One of the tubes is used to inject saliva into the foodstuff to begin liquefying it. The other tube is used like a soda straw to suck up liquid food.

While united by their type of mouthparts, members of various groups within the order Hemiptera differ as to other structures. The three main groups (suborders) of Hemiptera are Heteroptera, Auchenorrhyncha, and Sternorrhyncha. The first group, the Heteroptera, are called true bugs, and include such familiar groups as stink bugs, assassin bugs, and giant water bugs. The second, Auchenorrhyncha, is made up of cicadas, leafhoppers, treehoppers, froghoppers, and planthoppers. The last suborder, Sternorrhyncha, is home to aphids, whiteflies, scale insects, mealybugs, and plant lice.

The tarsi ("feet") are one structure that differs from suborder to suborder. In the Heteroptera the tarsi are two-segmented, while the tarsi of Auchenorrhyncha have three segments. Insects in Sternorrhyncha have tarsi that are one- or two-segmented.

While it is common for Hemiptera wings to have relatively few veins, other wing traits differ sharply within the order. For example, in Auchenorrhyncha and Sternorrhyncha the wings are membranous and fairly uniform. The forewings of Heteroptera, however, are thickened at the base, and typically opaque there, but are membranous and translucent at the apex. The word Heteroptera means "different wings," an apt descriptor for the true bugs.

Heteroptera nearly always hold their wings folded flat across the top of their body, and the wings characteristically form an "X" pattern when viewed from above. This helps differentiate true bugs from beetles, since beetles have wings that meet in a straight line down the center of their back. The Auchenorrhyncha and Sternorrhyncha tend to hold their wings roof-like over the back, though note that some groups are wingless.

Food. Because of the nature of their mouthparts, all Hemiptera have a liquid diet. Remember, though, that one of the two tubes injects digestive liquids into the food, so that solid food soon liquifies. A majority of hemipterans, including all the Auchenorrhyncha and Sternorrhyncha, feed on plants. Many Heteropterans are predatory, though, including, for example assassin bugs, the stink bug subfamily Asopinae, and most aquatic bugs.

Among the phytophagous (plant-feeding) species, some feed by liquifying plant tissues in the roots, stems, or leaves. Others, especially in Auchenorrhyncha and Sternorrhyncha, tap into sugary sap. Bugs ranging from leafhoppers to aphids excrete excess sugar and water from the anus, a "honeydew" that is in turn consumed by other creatures, most notably ants and other hymenopterans. Many ant colonies rely on honeydew, and worker ants carry droplets back to the nest. The hoppers, scale insects, and other honeydew producers benefit from this relationship, because the ants stay with them and drive away predators and prevent egg-laying by parasitoid wasps.

Figure 24 These *Paratrechina longicornis* ants are protecting a mealybug (Pseudococcidae), and later will feed on the honeydew the mealybug produces. Dominican Republic.

Reproduction. Hemiptera males and females find each other primarily by sight, though some species emit pheromones as attractants. Finding each other is made easier by some species' living only on one kind of plant. Bugs mate in a tail-to-tail position. Females then lay the eggs in a location where the nymphs will be able to find food easily. In a few species of Heteroptera, the female guards the eggs until hatching and may even stay with the young nymphs for a time after they hatch. The kind of parental care is fairly uncommon in class Insecta.

The most interesting reproduction is in suborder Sternorrhyncha. In some species in this group we find males inseminating immobile females, reproduction without sex, live births, and a practice called "traumatic insemination."

Because mature female scale insects in some families lack legs or wings, they must rely on males to find them for mating; mating takes places despite their immobility. The "traumatic insemination" in groups such as aphids consists of the male using his penis to pierce the abdominal wall of the female, since a vagina is not present. The sperm are collected by a sort of spongy structure within the female's abdomen. As for live birth, the females of some species hold the eggs within their abdomen until they hatch, then expel the hatchlings from their bodies.

It is the asexual (or parthenogenetic) reproduction that has most attracted the attention of researchers. Parthenogenesis is rare in the animal kingdom, but some species in Sternorrhyncha are wholly parthenogenetic. Many aphids, though, alternate between sexual and asexual reproduction. For example, early in the year a female may reproduce by generating clones of herself, all female, of course. Later in the year the female will mate with a male and produce offspring of both sexes.

Perhaps the most unusual sternorrhynchan reproduction is found in some members of the family Monophlebidae. Some hermaphroditic insects in this group, including the widespread cottony cushion scale *(Icerya purchasi)*, can produce eggs *and* fertilize them, since the insect has both ova and testes.

Development. Hemiptera are characterized by incomplete metamorphosis, or hemimetabolism. Thus, the hatchlings look like small adults except for their lack of wings. As they feed and grow, they must molt periodically as they outgrow their integument. Most bugs go through about five instars before the final molt to adulthood. Bug nymphs live lives much like those of the adults, in the same habitat, though without mating or flying.

Though Hemiptera are hemimetabolic, some Sternorrhycha stretch the definition by going through a resting phase, sometimes called a pupal stage. Some scale insects even produce a cocoon.

Dangers to Hemiptera. The primary danger in the life of hemipterans is predation. Birds, spiders, lizards, and fish are among the creatures willing to make a meal of a bug. Predacious insects, including other hemipterans, also prey upon this group.

Man-made chemicals take their toll, both in the form of insecticides and in the overall pollution of air, water, and soil by emissions from factory, farms, mines, and vehicles.

It appears bugs, hoppers, and whiteflies are less subject to infestation by parasites and parasitoids than some other groups such as Lepidoptera, but such infestations do occur. True bugs and hoppers are often victims of entomopathogenic fungi, which can both cover and penetrate their bodies, resulting in death.

Figure 25 Stink bug (Pentatomidae) hatchlings emerge from drum-shaped eggs and remain together for a time. A spider is lurking at lower left. Nicaragua.

Figure 26 The aposematic (warning) coloration of these leaf-footed bug nymphs *(Pachylis sp.* or *Thasus sp.)* may be more effective when the nymphs congregate together, making one big warning display. Nicaragua.

Species of bright red mites are sometimes seen on various kinds of hoppers. It is often hard to tell if such mites are feeding on the hopper's haemolymph or are simply taking advantage of transportation to a new location. Tachinid flies have been documented laying their eggs on adult leaf-footed bugs and other heteropterans. The fly larvae then slowly consume their host.

Figure 27 This planthopper (Flatidae) is a victim of entomopathogenic fungus that has covered much of its body. Puerto Rico.

Defenses. All heteropteran nymphs and adults have glands that produce noxious or repellant chemicals useful for defense against predators. Species in the other two suborders are often chemically protected, too—aphids, for example, can release an anti-predator glue from their cornicles (the paired tubes found at the top rear of their abdomens). Many heteropterans sport warning coloration in red or orange and black to discourage predators from attacking in the first place.

Some bugs are cryptically colored to blend in with their substrate, while others seek out protected places like crevices in bark. Some bugs, especially nymphs, resemble ants. Predators will often avoid attacking anything that looks like an ant, since ants are themselves fearsome predators.

Some species of bug in all three suborders have developed relationships with ants. Some live part-time or full-time in ant colonies and produce chemicals that cause the ants to accept them. More commonly, hoppers, scale insects, and their relatives feed out in the open while ants or other Hymenoptera stand guard and repel any attackers. As mentioned above, the bugs feed the Hymenoptera honeydew, which consists primarily of the amounts of sugar and water not needed by the bug for its own needs. Some Hymenoptera even behave like shepherds, moving their "flock" of hemipterans to better feeding locations or taking them into their own colonies during adverse weather.

Cicada nymphs in wet areas protect themselves from minor flooding by constructing chimneys that protect the underground burrow from water and mudslides. The chimney ensures the nymphs can exit their underground homes even if water covers the ground.

Relationship to humans. Heteroptera are one of the more important groups affecting human health. Bedbugs have tormented humans for millennia, but the long-term health effects of the bites tend to be minor. Persons with many bites may suffer from anemia due to blood loss. A more serious health threat to humans comes from kissing bugs, a type of assassin bug in subfamily Triatominae. The 127 Neotropical species in this subfamily can all transmit the protozoa *Trypanosoma cruzi*, which is the vector of Chagas disease. However, only the species that are regularly found in human dwellings are considered a real health risk; these include *Triatoma infestans* and *Rhodnius prolixus*. Chagas disease is a debilitating illness for which there is no vaccine and no cure.

Figure 28 Exuvia (molted "skin") of a cicada. The skin split along the back, and the teneral cicada exited through the resulting slit. Unfortunately, few Neotropical cicadas can be identified from the exuvia alone at present. Panama.

Many species in Hemiptera are agricultural pests. Miridae (plant bugs) and Cicadellidae (leafhoppers) are two of the more important offending families. The entire suborders Auchenorrhyncha and Sternorrhyncha harm plants, both crops and weed species. The harm comes from taking some of the plants' sap, and in many cases from the diseases the insect introduces into the plant. While predacious heteropterans sometimes prey on insect pests, they also prey on beneficial ones such as pollinators, and thus the group has a mixed record of benefit and harm. But most Hemipteran species have no effect on humans except as members of well-functioning ecosystems that benefit humankind.

Neotropical Specialties. Several taxa of Hemiptera are found only in the Neotropics. One such group is the true bug family Canopidae, made up of eight species all in the genus *Canopus*. These black bugs have the gloss of patent leather. They are shaped a bit like a high-domed lady beetle, and indeed members of the family are easily mistaken for beetles. The head is somewhat triangular, and in some species the eyes are red and the antennae banded.

One of the most interesting Neotropical species in Auchenorrhyncha is *Fulgora laternaria*, the so-called lanternfly or peanut-headed bug. There are several similar species in the genus. These bugs have a peanut-shaped extension to the front of the head that resembles the head of a snake or tiny caiman. This "face" may have the effect of frightening away would-be predators. Any startle reaction caused by the lizard-like head is compounded by a pair of eyespots on the hindwings that can be suddenly flashed at an attacker. The common name lanternfly refers to a mistaken notion of early naturalists that this species gives off light at night. Another mistaken belief is one widespread in Latin America today that these insects are capable of a venomous bite or sting. In fact, they are harmless. Other fulgorid planthoppers, and treehoppers too, boast bizarre structures and technicolor markings.

Figure 19 Two cicada chimneys in the Amazon basin of Brazil. The structures function to protect their builders from the effects of flooding.

Figure 30 These cotton stainers *(Dysdercus andreae)* are using their bright, aposematic coloration to warn predators of their chemical defenses. The bugs are clustered on top of a large seed pod, a food source. Martinique.

Gerridae: Gerrinae
Limnogonus sp. • Water Strider
Novo Airão, Amazonas, Brazil

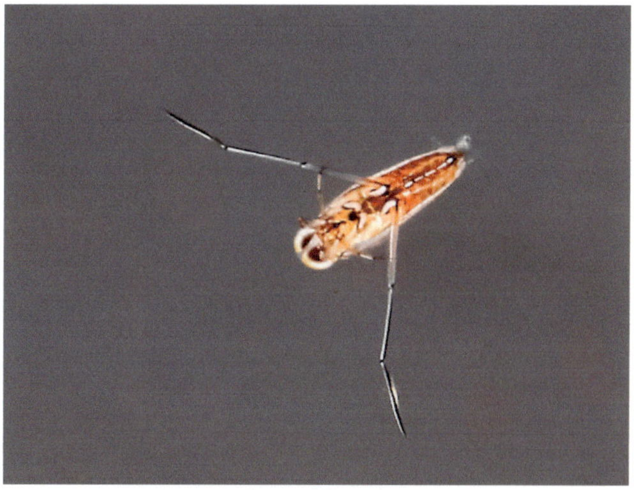

Notonectidae: Notonectidae
Notonecta sp. • Backswimmer
Parque Nacional Soberanía, Colón, Panama

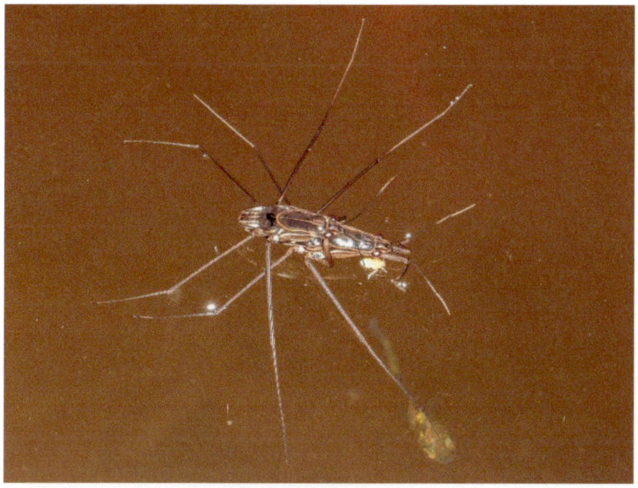

Gerridae: Gerrinae
Limnogonus sp. • Water Strider
Valle de Antón, Coclé, Panama

Gelastocoridae: Gelastocorinae
Gelastocoris sp. • Toad Bug
Parque Nacional Soberanía, Colón, Panama

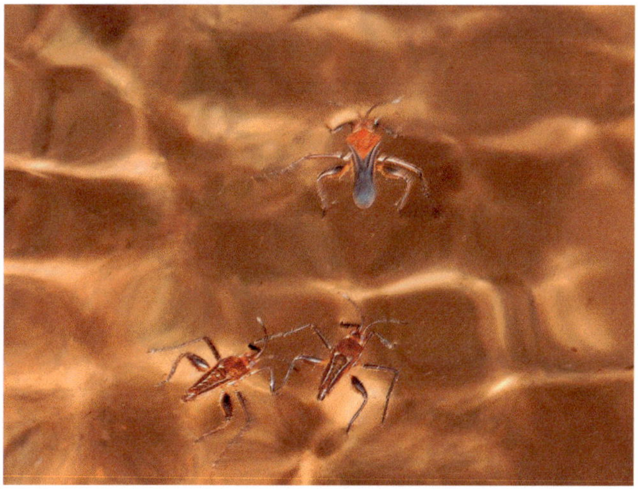

Veliidae: Rhagoveliinae
Rhagovelia sp. • Broad-shouldered Water Strider
Selva Negra, Matagalpa, Nicaragua

Gelastocoridae: Gelastocorinae
Gelastocoris sp. • Toad Bug
Parque Nacional Soberanía, Colón, Panama

Gelastocoridae: Gelastocorinae
Gelastocoris sp. • Toad Bug nymph
Colakreek, Para, Suriname

Miridae: Mirinae
Collaria scenica • Plant Bug
Along Rio de la Plata, Colonia, Uruguay

Gelastocoridae: Gelastocorinae
Gelastocoris sp. • Toad Bug nymph
Valle de Antón, Coclé, Panama

Miridae: Mirinae
Calocorisca sp. • Plant Bug
Boquete, Chiriquí, Panama

Miridae: Mirinae
Minytus nr. agrillaceus • Plant Bug
Montagne de Kaw, Roura, French Guiana

Miridae: Mirinae
Phytocoris sp. • Plant Bug
Monteverde, Puntarenas, Costa Rica

Miridae: Mirinae

Phytocoris sp. • Plant Bug

Boquete, Chiriquí, Panama

Miridae: Mirinae

Calondas fasciatus • Banded Calondas

Tuis, Cartago, Costa Rica

Miridae: Mirinae

Polymerus testaceipes • Plant Bug

Montagne de Kaw, Roura, French Guiana

Miridae: Mirinae

≈ Prepops sp. • Plant Bug

Filadelfia, Boquerón, Paraguay

Miridae: Mirinae

Taedia sp. • Plant Bug

Pico Bonito National Park lodge, Atlántida, Honduras

Miridae: Cylapinae

Fulvius sp. • Cylapine Plant Bug

Montagne de Kaw, Roura, French Guiana

Miridae: Cylapinae
Peltidocylapus sp. • Cylapine Plant Bug
Altos del Maria, Coclé, Panama

Miridae: Bryocorinae
Cyrtocapsus sp. • Bryocorine Plant Bug
Berg en Dal, Brokopondo, Suriname

Miridae: Deraeocorinae
Hyaliodocoris sp. • Hyaline Plant Bug
Selva Negra, Matagalpa, Nicaragua

Miridae: Orthotylinae
Jobertus chryselectrus • Orthotyline Plant Bug
Long Bay, Portland parish, Jamaica

Reduviidae: Ectrichodiinae
≈ Rhiginia sp. • Millipede Assassin
Shintuya Hot Springs, Madre de Dios, Peru

Reduviidae: Emesinae
Ghinallelia sp. • Thread-legged Assassin
Reserva Particular do Cristalino, Mato Grosso, Brazil

Reduviidae: Reduviinae
Zelurus cf. spinidorsis • Assassin Bug
Along the Rupununi River, Guyana

Reduviidae: Harpactorinae
Pselliopus sp. • Bracelet Assassin Bug
El Hato, Sacatepéquez, Guatemala

Reduviidae: Reduviinae
Zelurus sp. • Assassin Bug
Tuis, Cartago, Costa Rica

Reduviidae: Harpactorinae
Pselliopus nr. cinctus • Bracelet Assassin Bug
El Hato, Sacatepéquez, Guatemala

Reduviidae: Reduviinae
Zelurus salyavatoides • Assassin Bug
Berg en Dal, Brokopondo, Suriname

Reduviidae: Harpactorinae
Castolus sp. • Harpactorine Assassin Bug
Santo Domingo de los Colorados, Pichincha, Ecuador

Reduviidae: Harpactorinae
Castolus tricolor • Tri-colored Assassin Bug
Monteverde, Puntarenas, Costa Rica

Reduviidae: Apiomerinae
Apiomerus geniculatus • Bee Assassin
Along the Rupununi River, Guyana

Reduviidae: Harpactorinae
Montina sp. • Harpactorine Assassin Bug
Berg en Dal, Brokopondo, Suriname

Reduviidae: Harpactorinae
Apiomerus lanipes • Bee Assassin
Along the Rupununi River, Guyana

Reduviidae: Harpactorinae
Apiomerus cf. venustus • Bee Assassin
Novo Airão, Amazonas, Brazil

Reduviidae: Harpactorinae
Notocyrtus sp. • Bee-mimic Assassin
Berg en Dal, Brokopondo, Suriname

Reduviidae: Apiomerinae
Micrauchenus lineola • Termite Assassin
Rewa village, Upper Takutu-Upper Essequibo, Guyana

Reduviidae: Harpactorinae
Heniartes sp. • Harpactorine Assassin Bug
Colakreek, Para, Suriname

Reduviidae: Harpactorinae
Agriocoris flavipes • Harpactorine Assassin Bug
Montsinéry, Montsinéry-Tonnegrande, French Guiana

Reduviidae: Harpactorinae
Harpactorinae, possibly Heza sp. • Assassin Bug nymph
Berg en Dal, Brokopondo, Suriname

Reduviidae: Harpactorinae
Heniartes sp. • Harpactorine Assassin Bug
Montsinéry, Montsinéry-Tonnegrande, French Guiana

Reduviidae: Harpactorinae
Heza sp. • Harpactorine Assassin Bug
Along the Rewa River, Guyana

Reduviidae: Harpactorinae
Repipta antica • Assassin Bug
Along the Rewa River, Guyana

Reduviidae: Harpactorinae
Repipta flavicans • Yellowish Repipta
Along the Rewa River, Guyana

Reduviidae: Harpactorinae
Repipta lepidula • Small-scaled Repipta
Réserve naturelle Trésor, Roura, French Guiana

Reduviidae: Harpactorinae
Ricolla pallidinervis • Pale-veined Ricolla
Tuis, Cartago, Costa Rica

Reduviidae: Harpactorinae
Ricolla quadrispinosa • Four-spined Ricolla
Berg en Dal, Brokopondo, Suriname

Reduviidae: Harpactorinae
Zelus longipes • Long-legged Zelus
Crooked Tree preserve, Belize district, Belize

Reduviidae: Harpactorinae
≈ Zelus sp. • Assassin nymph
Montsinéry, Montsinéry-Tonnegrande, French Guiana

Reduviidae: Bactrodinae
Bactrodes cf. femoratus • Bactrodine Ambush Bug
Sentier de Bagne des Annamites, French Guiana

Reduviidae: Harpactorinae
Zelus sp. • Assassin Bug
Selva Negra, Matagalpa, Nicaragua

Reduviidae: Bactrodinae
Bactrodes spinulosus • Spiny Bactrodes
El Hato, Sacatepéquez, Guatemala

Reduviidae: Harpactorinae
Rhiginia sp. • Assassin Bug
Shintuya Hot Springs, Madre de Dios, Peru

Reduviidae: Stenopodainae
Stenopoda sp. • Assassin Bug
Parque Nacional Pico Bonito, lodge, Atlántida, Honduras

Reduviidae: Peiratinae
≈ Rasahus sp. • Assassin Bug nymph
Berg en Dal, Brokopondo, Suriname

Aradidae: Mezirinae
Aphleboderrhis pubescens • Downy Flat Bug
Along the Rewa River, Guyana

Reduviidae: Phymatinae
Lophoscutus sp. • Ambush Bug
Tuis, Cartago, Costa Rica

Aradidae: Mezirinae
Hesus acuminatus • Pointed Flat Bug
Reserva Particular do Cristalino, Mato Grosso, Brazil

Tingidae: Tinginae
Corythucha cf. ciliate • Sycamore Lace Bug
Colonia del Sacramento, Colonia, Uruguay

Aradidae: Mezirinae
Hesus cordatus • Flat Bug
Tuis, Cartago, Costa Rica

Aradidae: Mezirinae
Hesus humeralis • Shouldered Flat Bug
Along the Rio Negro, Amazonas, Brazil

Aradidae: Mezirinae
Phyllotingis eximia • Exceptional Flat Bug
Manu National Park, Madre de Dios Department, Peru

Aradidae: Mezirinae
Placogenys sp. • Flat Bug
Along the Rewa River, Guyana

Berytidae: Metacanthinae
Jalysus sp. • Stilt Bug
El Hato, Sacatepéquez, Guatemala

Berytidae: Metacanthinae
≈ Metacanthus sp. • Stilt Bug
Paracas, Ica, Peru

Geocoridae: Geocorinae
Geocoris sp. • Big-eyed Bug nymph
Paracas, Ica, Peru

Lygaeidae: Lygaeinae
Oncopeltus fasciatus • Large Milkweed Bug
Bosque Estatal Rio Abajo, Puerto Rico

Lygaeidae: Lygaeinae
Lygaeus nr. kalmii • Seed Bug
Savaneta, Savaneta, Aruba

Lygaeidae: Lygaeinae
Latochrimnus sp. • Seed Bug
Monteverde, Puntarenas, Costa Rica

Colobathristidae
Colobathristidae • Colobathristid Bug
Lake Machuwasi, Manú Natl. Park, Madre de Dios, Peru

Lygaeidae: Lygaeinae
Lygaeus froeschneri • Froeschner's Seed Bug
Lake Titicaca shore, Yumani, Bolivia

Pachygronthidae: Pachygronthinae
≈ Oedancala sp. • Pachygronthine Seed Bug
Lagunas de Aruza, Darién, Panama

Rhyparochromidae: Rhyparochrominae
Neopamera sp. • Rhyparochromid Seed Bug
Arima Valley, Trinidad

Rhyparochromidae: Rhyparochrominae
≈ Ozophora sp. • Rhyparochromid Seed Bug
Réserve naturelle Trésor, Roura, French Guiana

Rhyparochromidae: Rhyparochrominae
Neopamera costalis • Rhyparochromid Seed Bug
Playa Los Angeles, Magdalena, Colombia

Rhyparochromidae: Rhyparochrominae
Pephysena sp. • Rhyparochromid Seed Bug
Reserva Particular do Cristalino, Mato Grosso, Brazil

Rhyparochromidae: Rhyparochrominae
Myodocha sp. • Long-necked Seed Bug nymph
Tortuguero National Park, Limón, Costa Rica

Pyrrhocoridae: Pyrrhocorinae
Dysdercus andreae • St. Andrew's Cotton Stainer
Long Bay, Portland parish, Jamaica

Pyrrhocoridae: Pyrrhocorinae
Dysdercus sp. • Cotton Stainer
Novo Airão, Amazonas, Brazil

Pyrrhocoridae: Pyrrhocorinae
Dysdercus obliquus • Cotton Stainer
Santa Elena, Puntarenas, Costa Rica

Pyrrhocoridae: Pyrrhocorinae
Dysdercus mimulus • Cotton Stainer
Selva Negra, Matagalpa, Nicaragua

Largidae: Larginae
Fibrenus globicollis • Hump-collared Stainer
Leona Vicario, Quintana Roo, Mexico

Pyrrhocoridae: Pyrrhocorinae
Dysdercus obscuratus • Cotton Stainer
Santa Elena, Puntarenas, Costa Rica

Largidae: Larginae
Largus cf. maculatus • Bordered Plant Bug
Monteverde, Puntarenas, Costa Rica

Largidae: Larginae
Largus rufipennis • Red-winged Largus
Médanos de Punta Gorda, Colonia, Uruguay

Alydidae: Alydinae
Hyalymenus sp. • Broad-headed Bug
Pico Bonito National Park, Atlántida, Honduras

Largidae: Larginae
Stenomacra marginella • Bordered Plant Bug
El Hato, Sacatepéquez, Guatemala

Alydidae: Alydinae
Hyalymenus sp. • Broad-headed Bug nymph
Pico Bonito National Park, Atlántida, Honduras

Largidae: Larginae
Theraneis nr. pulchra • Bordered Plant Bug
Tuis, Cartago, Costa Rica

Alydidae: Alydinae
Hyalymenus sp. • Broad-headed Bug nymph
Pico Bonito National Park, Atlántida, Honduras

Alydidae: Alydinae

Hyalymenus vespiformis • Wasp-mimic Broad-head Bug

Along the Rupununi River, Guyana

Alydidae: Alydinae

Apidaurus longistylus • Broad-headed Bug

Montagne de Kaw, Roura, French Guiana

Alydidae: Alydinae

Neomegalotomus parvus • Broad-headed Bug

Novo Airão, Amazonas, Brazil

Alydidae: Alydinae

Apidaurus cf. consperus • Broad-headed Bug nymph

12 km S of Filadelfia, Boquerón, Paraguay

Alydidae: Alydinae

Neomegalotomus sp. • Broad-headed Bug

Montagne de Kaw, Roura, French Guiana

Alydidae: Alydinae

Burtinus notatipennis • Spot-winged Broad-headed Bug

El Hato, Sacatepéquez, Guatemala

Alydidae: Micrelytrinae
Stenocoris sp. • Rice Bug
Novo Airão, Amazonas, Brazil

Coreidae: Meropachydinae
Peranthus sp. or Hirilcus sp. • Leaf-footed Bug
Reserva Particular do Cristalino, Mato Grosso, Brazil

Coreidae: Meropachydinae
Lycambes sp. • Leaf-footed Bug
Selva Negra, Matagalpa, Nicaragua

Coreidae: Coreinae
Catorhintha cf. guttula • Leaf-footed Bug
Novo Airão, Amazonas, Brazil

Coreidae: Meropachydinae
≈ Paralycambes sp. • Leaf-footed Bug
Reserva Particular do Cristalino, Mato Grosso, Brazil

Coreidae: Coreinae
Chariesterus sp. • Leaf-footed Bug
Lagunas de Aruza, Darién, Panama

Coreidae: Coreinae
Anisoscelis affinis • Allied Anisoscelis
Pico Bonito National Park, Atlántida, Honduras

Coreidae: Coreinae
Anisoscelini • Leaf-footed Bug nymph
Parque Nacional Pico Bonito, Atlántida, Honduras

Coreidae: Coreinae
Anisoscelis alipes • Wing-footed Anisoscelis
Sansón, Darién, Panama

Coreidae: Coreinae
Leptoglossus gonagra • Leaf-footed Bug
Maniche, département du Sud, Haiti

Coreidae: Coreinae
Diactor bilineatus • Two-lined Diactor
Sansón, Darién, Panama

Coreidae: Coreinae
Leptoglossus zonatus • Belted Leptoglossus
Reserva Particular do Cristalino, Mato Grosso, Brazil

Coreidae: Coreinae
Leptoscelis quadrisignatus • Four-marked Leptoscelis
Tuis, Cartago, Costa Rica

Coreidae: Coreinae
Petalops sp. • Leaf-footed Bug
Novo Airão, Amazonas, Brazil

Coreidae: Coreinae
Leptoscelis tricolor • Heliconia Bug
Valle de Antón, Coclé, Panama

Coreidae: Coreinae
Paryphes nr. jucundus • Leaf-footed Bug
Manú National Park, Madre de Dios Department, Peru

Coreidae: Coreinae
Cnemomis sp. • Leaf-footed Bug
Novo Airão, Amazonas, Brazil

Coreidae: Coreinae
Petalops thoracicus • Leaf-footed Bug
Along the Rewa River, Guyana

Coreidae: Coreinae
Zoreva sp. • Leaffooted Bug
Berg en Dal, Brokopondo, Suriname

Coreidae: Coreinae
Melucha sp. • Leaf-footed Bug
Montsinéry, Montsinéry-Tonnegrande, French Guiana

Coreidae: Coreinae
Sagotylus confluentus • Confluent Leaffooted Bug
Long Bay, Portland parish, Jamaica

Coreidae: Coreinae
Stenoscelidea tibialis • Leaf-footed Bug
Manú National Park, Madre de Dios, Peru

Coreidae: Coreinae
Sagotylus confluentus • Leaffooted Bug nymph
Long Bay, Portland parish, Jamaica

Coreidae: Coreinae
Hypselonotus concinnus • Leaf-footed Bug
Selva Negra, Matagalpa, Nicaragua

Coreidae: Coreinae
Hypselonotus interruptus • Leaf-footed Bug
Santo Domingo de los Colorados, Pichincha, Ecuador

Coreidae: Coreinae
Nematopus cf. caicarensis • Leaf-footed Bug
Novo Airão, Amazonas, Brazil

Coreidae: Coreinae
Anasa madida or A. peregrina • Squash Bug
Monteverde, Puntarenas, Costa Rica

Coreidae: Coreinae
Nematopus sp. • Leaf-footed Bug
Reserva Particular do Cristalino, Mato Grosso, Brazil

Coreidae: Coreinae
Nematopus indus • Leaf-footed Bug
Berg en Dal, Brokopondo, Suriname

Coreidae: Coreinae
Pachylis phraraonis • Leaf-footed Bug
Reserva Particular do Cristalino, Mato Grosso, Brazil

Coreidae: Coreinae
Piezogaster rubropictus • Red-marked Piezogaster
Manú Road, Cusco department, Peru

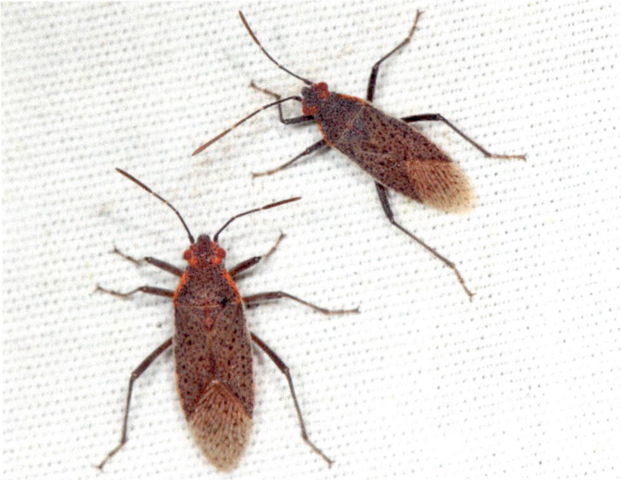

Rhopalidae: Serinethinae
Jadera cf. coturnix • Scentless Plant Bug
Pico Bonito National Park lodge, Atlántida, Honduras

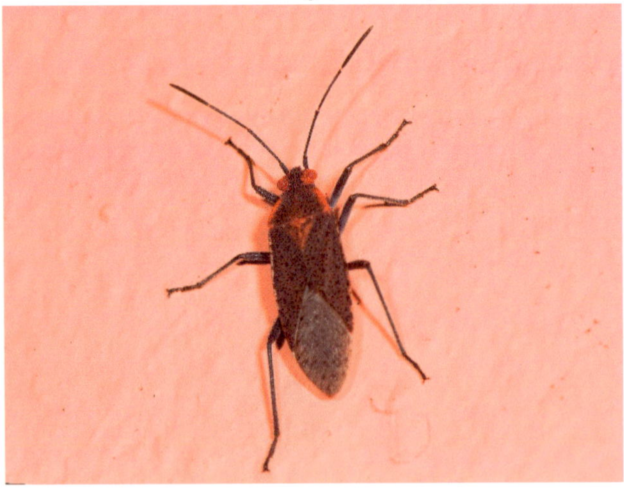

Rhopalidae: Serinethinae
Jadera sanguinolenta • Scentless Plant Bug
Long Bay, Portland parish, Jamaica

Cydnidae: Cydninae
Pangaeus sp. • Burrowing Bug
Along the Rupununi River, Guyana

Cydnidae: Cydninae
Pangaeus sp. • Burrowing Bug
Altos del Maria, Coclé, Panama

Pentatomidae: Pentatominae
Fecelia sp. • Stink Bug
Rabo de Gato, Sierra de Bahoruco, Dominican Republic

Pentatomidae: Pentatominae
Loxa sp. • Stink Bug
Tuis, Cartago, Costa Rica

Pentatomidae: Pentatominae
Euschistus sp. • Stink Bug
Parque Nacional Tayrona, Magdalena, Colombia

Pentatomidae: Pentatominae
Proxys albopunctulatus / P. victor • Stink Bug
Colakreek, Para, Suriname

Pentatomidae: Pentatominae
Euschistus sp. • Stink Bug
Manaus, Amazonas, Brazil

Pentatomidae: Pentatominae
Euschistus cf. heros • Stink Bug
Along the Rupununi River, Guyana

Pentatomidae: Pentatominae
Mormidea cf. ypsilon • Ypsilon Stink Bug
Along the Rupununi River, Guyana

Pentatomidae: Pentatominae
Sibaria sp. • Stink Bug
Along the Rupununi River, Guyana

Pentatomidae: Pentatominae
Banasa cf. zeteki • Stink Bug
Tuis, Cartago, Costa Rica

Pentatomidae: Pentatominae
Banasa sp. • Stink Bug
Rewa village, Upper Takutu-Upper Essequibo, Guyana

Pentatomidae: Pentatominae
Murgantia cf. bifasciata • Two-banded Stink Bug
Manú Road, Cusco Department, Peru

Pentatomidae: Pentatominae
Banasa lenticularis • Stink Bug
Montagne de Kaw, Roura, French Guiana

Pentatomidae: Pentatominae
Pellaea stictica • Stink Bug
Tuis, Cartago, Costa Rica

Pentatomidae: Discocephalinae
Antiteuchus sp. • Disk-headed Stink Bug
Manaus, Amazonas, Brazil

Pentatomidae: Discocephalinae
Pelidnocoris stalii • Disk-headed Stink Bug
Tuis, Cartago, Costa Rica

Pentatomidae: Discocephalinae
Antiteuchus sp. • Disk-headed Stink Bug
Novo Airão, Amazonas, Brazil

Pentatomidae: Discocephalinae
Schraderiellus cf. cinctus • Stink Bug
Tuis, Cartago, Costa Rica

Pentatomidae: Discocephalinae
Cataulax pudens • Disk-headed Stink Bug
Parque Nacional Pico Bonito, lodge, Atlántida, Honduras

Pentatomidae: Pentatominae
Chlorocoris distinctus • Distinctive Stink Bug
Tuis, Cartago, Costa Rica

Pentatomidae: Edessinae
Brachystethus rubromaculatus • Red-spotted Stink Bug
Pico Bonito National Park, Atlántida, Honduras

Pentatomidae: Edessinae
Edessa rufomarginata • Red-margined Stink Bug
Novo Airão, Amazonas, Brazil

Pentatomidae: Edessinae
Brachystethus cf. cribus • Edessine Stink Bug
Reserva Particular do Cristalino, Mato Grosso, Brazil

Pentatomidae: Edessinae
Edessa nr. bifida • Edessine Stink Bug
Berg en Dal, Brokopondo, Suriname

Pentatomidae: Edessinae
Edessa nr. pictiventris • Stink Bug
Pico Bonito National Park lodge, Atlántida, Honduras

Pentatomidae: Edessinae
Edessa guyanensis • Guyana Stink Bug
Montagne de Kaw, Roura, French Guiana

Pentatomidae: Edessinae
Edessa cf. costalis • Edessine Stink Bug
Berg en Dal, Brokopondo, Suriname

Pentatomidae: Edessinae
Edessa nr. urus • Edessine Stink Bug
Reserva Particular do Cristalino, Mato Grosso, Brazil

Pentatomidae: Edessinae
Edessa scutellata • Edessine Stink Bug
Berg en Dal, Brokopondo, Suriname

Pentatomidae: Edessinae
Edessa sp. • Edessine Stink Bug
Monteverde, Puntarenas, Costa Rica

Pentatomidae: Edessinae
Edessa cf. cortezi – rixosa group • Edessine Stink Bug
Leona Vicario, Quintana Roo, Mexico

Pentatomidae: Edessinae
Edessa sp. • Edessine Stink Bug
Reserva Particular do Cristalino, Mato Grosso, Brazil

Pentatomidae: Edessinae
Peromatus sp. • Edessine Stink Bug nymph
Along the Rupununi River, Guyana

Pentatomidae: Asopinae
Euthyrhynchus floridanus • Florida Predatory Stink Bug
Selva Negra, Matagalpa, Nicaragua

Pentatomidae: Edessinae
Peromatus cf. notatus • Edessine Stink Bug
Along the Rupununi River, Guyana

Pentatomidae: Asopinae
Podisus aenescens • Predatory Stink Bug
Berg en Dal, Brokopondo, Suriname

Pentatomidae: Asopinae
Apateticus lineolatus • Predatory Stink Bug
Monteverde, Puntarenas, Costa Rica

Scutelleridae: Pachycorinae
Symphylus sp. • Shield Bug
Tuis, Cartago, Costa Rica

Scutelleridae: Pachycorinae
Symphylus sp. • Shield Bug
Parque Nacional Soberanía, Colón, Panama

Canopidae: Canopinae
Canopus sp. • Canopid Bug
Along the Rewa River, Guyana

Scutelleridae: Pachycorinae
Pachycoris torridus • Physic Nut Shield Bug
Reserva Particular do Cristalino, Mato Grosso, Brazil

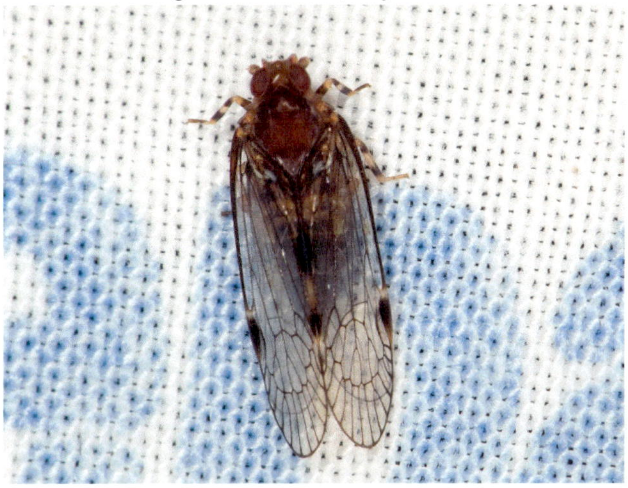

Cixiidae: Cixiinae
Cixius sp. • Cixiid Planthopper
Montagne de Kaw, Roura, French Guiana

Corimelaenidae: Thyreocorinae
Galgupha sp. • Ebony Bug
Montsinéry, Montsinéry-Tonnegrande, French Guiana

Cixiidae: Cixiinae
Cixiinae • Cixiine Planthopper
Colakreek, Para, Suriname

Cixiidae: Cixiinae
Pintalia sp. • Cixiid Planthopper
Along the Rio Negro, Amazonas, Brazil

Cixiidae: Cixiinae
Pintaliini • Cixiid Planthopper
Tuis, Cartago, Costa Rica

Cixiidae: Cixiinae
Pintalia sp. • Cixiid Planthopper
Monteverde, Puntarenas, Costa Rica

Cixiidae: Cixiinae
≈ Melanoliarus sp. • Cixiid Planthopper
Along the Rio Negro, Amazonas, Brazil

Cixiidae: Cixiinae
Pintaliini • Cixiid Planthopper
Along the Rewa River, Guyana

Cixiidae: Cixiinae
≈ Melanoliarus sp. • Cixiid Planthopper
Along the Rupununi River, Guayana

Cixiidae: Cixiinae
Near Muirolonia sp. • Cixiid Planthopper
Nanegalito, Tandayapa Valley, Ecuador

Cixiidae: Bothriocerinae
Bothriocera sp. • Cixiid Planthopper
Long Bay, Portland parish, Jamaica

Cixiidae: Cixiinae
Oecleus sp. • Cixiid Planthopper
Montsinéry, Montsinéry-Tonnegrande, French Guiana

Cixiidae: Bothriocerinae
Bothriocera sp. • Cixiid Planthopper
Altos del Maria, Coclé, Panama

Cixiidae: Bothriocerinae
Bothriocera sp. • Cixiid Planthopper
Roura-Kaw Road, Roura, French Guiana

Derbidae: Cedusinae
Cedusa maculata • Derbid Planthopper
Suchitoto, Cuscatlán, El Salvador

Derbidae: Derbinae
≈ Cenchreini • Derbid Planthopper
Colakreek, Para, Suriname

Derbidae: Derbinae
≈ Mysidia sp. • Moth Derbid
Along the Rio Negro, Amazonas, Brazil

Derbidae: Derbinae
Omolicna sp. • Derbid Planthopper
Rewa village, Upper Takutu-Upper Essequibo, Guyana

Derbidae: Derbinae
Mysidia sp. • Moth Derbid
Sentier de Bagne des Annamites, French Guiana

Derbidae: Derbinae
Omolicna sp. • Derbid Planthopper
Monteverde, Puntarenas, Costa Rica

Derbidae: Derbinae
Mysidini • Moth Derbid
Along the Rio Negro, Amazonas, Brazil

Delphacidae: Delphacinae
Phrictopyga contorta • Delphacid Planthopper
Montagne de Kaw, Roura, French Guiana

Delphacidae: Asiracinae
Copicerus cf. irroratus • Delphacid Planthopper
Selva Negra, Matagalpa, Nicaragua

Delphacidae: Delphacinae
Neomalaxa cf. flava • Delphacid Planthopper
Colakreek, Para, Suriname

Delphacidae: Asiracinae
Ugyops sp. • Delphacid Planthopper
Tuis, Cartago, Costa Rica

Delphacidae: Delphacinae
Sogatella kolophon • Delphacid Planthopper
Presqu'île de la Caravelle, Martinique

Delphacidae: Asiracinae
Ugyops sp. • Delphacid Planthopper
Along Rio Negro, Amazonas, Brazil

Delphacidae: Asiracinae
Ugyops cf. occidentalis • Delphacid Planthopper
Bosque Estatal Rio Abajo, Puerto Rico

Flatidae: Flatinae
≈ Flatormenis sp. • Flatid Planthopper
Leona Vicario, Quintana Roo, Mexico

Flatidae: Flatinae
Poekilloptera phalaenoides • Flatid Planthopper
Montagne de Kaw, Roura, French Guiana

Flatidae: Flatinae
Ormenoides sp. • Flatid Planthopper
Novo Airão, Amazonas, Brazil

Flatidae: Flatinae
Byllis sp. • Flatid Planthopper
Puerto Madryn, Chubut, Argentina

Achilidae: Achilinae
Plectoderini • Achilid Planthopper
Colonia del Sacramento, Colonia, Uruguay

Achilidae: Achilinae
Plectoderes collaris • Collared Achilid
Along the Rupununi River, Guyana

Fulgoroidea
≈ Dictyopharidae • 3rd or 4th instar planthopper nymph
Crooked Tree Preserve, Belize district, Belize

Achilidae: Achilinae
≈ Sevia sp. • Achilid Planthopper
Along the Rio Negro, Amazonas, Brazil

Dictyopharidae
Dictyopharidae • late instar Dictyopharid nymph
Berg en Dal, Brokopondo, Suriname

Achilidae: Achilinae
Achilini • Achilid Planthopper
Berg en Dal, Brokopondo, Suriname

Fulgoridae: Aphaeninae
Copidocephala guttata • Fulgorid Planthopper
Pico Bonito National Park lodge, Atlántida, Honduras

Fulgoridae: Aphaeninae
Copidocephala nr. guttata • Fulgorid Planthopper
Lake Machuwasi, Madre de Dios Department, Peru

Fulgoridae: Fulgorinae
Diareusa sp. • Fulgorid Planthopper
Valle de Antón, Coclé, Panama

Fulgoridae: Amyclinae
Enchophora sanguinea • Red-eyed Fulgorid
Tuis, Cartago, Costa Rica

Fulgoridae: Fulgorinae
≈ Phrictus sp. • Fulgorid Planthopper nymph
Reserva Particular do Cristalino, Mato Grosso, Brazil

Fulgoridae: Fulgorinae
Diareusa sp. • Fulgorid Planthopper
Berg en Dal, Brokopondo, Suriname

Fulgoridae: Poiocerinae
Aracynthus sanguineus • Blood-stained Fulgorid
Berg en Dal, Brokopondo, Suriname

Fulgoridae: Poiocerinae
Calyptoproctus sp • Fulgorid Planthopper
Leona Vicario, Quintana Roo, Mexico

Fulgoridae: Cladodipterinae
Diacira cf. obliquata • Fulgorid Planthopper
Along the Rupununi River, Guyana

Fulgoridae: Poiocerinae
≈ Poblicia sp. • Fulgorid Planthopper
Pico Bonito National Park, Atlántida, Honduras

Tropiduchidae
Amapala ornata • Tropiduchid Planthopper
Pico Bonito National Park, Atlántida, Honduras

Fulgoridae: Phenacinae
Pterodictya reticulata • Fulgorid Planthopper
Along the Rupununi River, Guyana

Nogodinidae: Nogodininae
Biolleyana pictifrons • Net-winged Planthopper
Pico Bonito National Park, Atlántida, Honduras

Nogodinidae: Nogodininae
Biolleyana costalis • Net-winged Planthopper
Tuis, Cartago, Costa Rica

Nogodinidae: Nogodininae
Bladina sp. • Net-winged Planthopper
Parque Nacional Tayrona, Magdalena, Colombia

Nogodinidae: Nogodininae
Vutina sp. • Net-winged Planthopper
Reserva Particular do Cristalino, Mato Grosso, Brazil

Dictyopharidae: Dictyopharinae
≈ Lappida sp. • Dictyopharid Planthopper
Nanegalito, Tandayapa Valley, Ecuador

Nogodinidae: Nogodininae
Bladina sp. • Net-winged Planthopper
Along the Rupununi River, Guyana

Dictyopharidae: Dictyopharinae
Lappida sp. • Dictyopharid Planthopper
Montagne de Kaw, Roura, French Guiana

Dictyopharidae: Dictyopharinae
≈ Hyalodictyon sp. • Dictyopharid Planthopper
Along the Rewa River, Guyana

Dictyopharidae: Dictyopharinae
Taosa sp. • Dictyopharid Planthopper
Berg en Dal, Brokopondo, Suriname

Dictyopharidae: Dictyopharinae
Taosa herbida • Dictyopharid Planthopper
Parque Nacional Tayrona, Magdalena, Colombia

Dictyopharidae: Dictyopharinae
Nersia cf. florens • Dictyopharid Planthopper
Suchitoto, Cuscatlán, El Salvador

Dictyopharidae: Dictyopharinae
Taosa sp. • Dictyopharid Planthopper
Playa Los Angeles, Magdalena, Colombia

Issidae: Issinae
≈ Picumna sp. • Issid Planthopper
Along the Rewa River, Guyana

Issidae: Issinae
≈ Thionia sp. • Issid Planthopper nymph
Novo Airão, Amazonas, Brazil

Acanaloniidae
Acanalonia nr. conica • Acanaloniid Planthopper
Tuis, Cartago, Costa Rica

Issidae
Issidae • Issid planthopper nymph
Tuis, Cartago, Costa Rica

Cercopidae: Ischnorhininae
Catrimania sp. • Froghopper
Berg en Dal, Brokopondo, Suriname

Acanaloniidae: Acanaloniinae
≈ Acanalonia sp. • Acanaloniid Planthopper
Rewa village, Upper Takutu-Upper Essequibo, Guyana

Cercopidae: Ischnorhininae
Laccogrypota valida • Froghopper
Montagne de Kaw, Roura, French Guiana

Cercopidae: Ischnorhininae
Maxantonia lineola • Froghopper
Montagne de Kaw, Roura, French Guiana

Cercopidae: Ischnorhininae
Ocoaxo septemnotatus •Seven-spotted Spittlebug
Boquete, Chiriquí, Panama

Cercopidae: Ischnorhininae
Monecphora maculipennis • Spot-winged Monecphora
Pico Bonito National Park lodge, Atlántida, Honduras

Cercopidae: Ischnorhininae
Prosapia bicincta • Two-belted Froghopper
Selva Negra, Matagalpa, Nicaragua

Cercopidae: Ischnorhininae
Ocoaxo lineatus • Froghopper
Pico Bonito National Park lodge, Atlántida, Honduras

Cercopidae: Ischnorhininae
Sinopia signata • Marked Froghopper
Montagne de Kaw, Roura, French Guiana

Cercopidae: Ischnorhininae
Sphenorhina conspicua • Consipcuous Froghopper
Tuis, Cartago, Costa Rica

Cercopidae: Ischnorhininae
Iphirhina perfecta • Perfect Froghopper
Altos del Maria, Coclé, Panama

Cercopidae: Ischnorhininae
Sphenorhina hebes • Froghopper
Manú Road, Cusco department, Peru

Cercopidae: Ischnorhininae
Iphirhina sp. • Froghopper
Selva Negra, Matagalpa, Nicaragua

Cercopidae: Ischnorhininae
Sphenorhina latifascia or S. nigricepha • Froghopper
Valle de Antón, Coclé, Panama

Cercopidae: Ischnorhininae
Iphirhina sp. • Froghopper
Altos del Maria, Coclé, Panama

Cercopidae: Ischnorhininae
Mahanarva costaricensis • Costa Rican Mahanarva
Tuis, Cartago, Costa Rica

Cercopidae: Ischnorhininae
Zulia cf. pubescens • Pubescent Froghopper
Berg en Dal, Brokopondo, Suriname

Clastopteridae: Clastopterinae
Clastoptera sp. • Clastopterid Froghopper
Selva Negra, Matagalpa, Nicaragua

Cicadellidae: Aphrodinae
Xestocephalus sp. • Aphrodine Leafhopper
Nanegalito, Tandayapa Valley, Pichincha, Ecuador

Cicadellidae: Megophthalminae
≈ Agalliota sp. • Leafhopper
El Hato, Sacatepéquez, Guatemala

Cicadellidae: Idiocerinae
≈ Idiocerinae • Leafhopper
Montagne de Kaw, Roura, French Guiana

Cicadellidae: Gyponinae
Ponana sp. • Gyponine Leafhopper
Pico Bonito National Park lodge, Atlántida, Honduras

Cicadellidae: Gyponinae
Scaris sp. • Gyponine Leafhopper
Montagne de Kaw, Roura, French Guiana

Cicadellidae: Gyponinae
Ponana sp. • Gyponine Leafhopper
Leona Vicario, Quintana Roo, Mexico

Cicadellidae: Coelidiinae
Coelidiinae • Coelidiine Leafhopper
Montsinéry, Montsinéry-Tonnegrande, French Guiana

Cicadellidae: Gyponinae
Scaris sp. • Gyponine Leafhopper
Along the Rewa River, Guyana

Cicadellidae: Coelidiinae
Teruliini• Coelidiine Leafhopper
Naranjo, Alajuela, Costa Rica

Cicadellidae: Coelidiinae
Coelidiinae • Coelidiine Leafhopper
Montsinéry, Montsinéry-Tonnegrande, French Guiana

Cicadellidae: Coelidiinae
Coelidiinae • Coelidiine Leafhopper
Rewa village, Upper Takutu-Upper Essequibo, Guyana

Cicadellidae: Coelidiinae
Coelidiinae • Coelidiine Leafhopper
El Hato, Sacatepéquez, Guatemala

Cicadellidae: Coelidiinae
Coelidiinae • Coelidiine Leafhopper nymph
El Hato, Sacatepéquez, Guatemala

Cicadellidae: Coelidiinae
Coelidiinae • Coelidiine Leafhopper
Along the Rupununi River, Guyana

Cicadellidae: Cicadellinae
Cicadellini • Sharpshooter Nymph
Along the Rio Negro, Amazonas, Brazil

Cicadellidae: Cicadellinae
Proconiini • Proconiini Leafhopper nymph
Suchitoto, Cuscatlán, El Salvador

Cicadellidae: Cicadellinae
Beirneola anita • Sharpshooter
Valle de Antón, Coclé, Panama

Cicadellidae: Cicadellinae
Baleja rufofasciata • Red-banded Sharpshooter
Tuis, Cartago, Costa Rica

Cicadellidae: Cicadellinae
Beirneola sp. • Sharpshooter
Nanegalito, Tandayapa Valley, Ecuador

Cicadellidae: Cicadellinae
Baleja flavoguttata • Yellow-spotted Sharpshooter
Valle de Antón, Coclé, Panama

Cicadellidae: Cicadellinae
Hadria nr. cubana • Sharpshooter
Maniche, département du Sud, Haiti

Cicadellidae: Cicadellinae
Onega cf. freytagi • Sharpshooter
Nanegalito, Tandayapa Valley, Ecuador

Cicadellidae: Cicadellinae
Oragua insipida • Insipid Sharpshooter
Berg en Dal, Brokopondo, Suriname

Cicadellidae: Cicadellinae
Oragua cf. osborni • Osborn's Sharpshooter
Rewa village, Upper Takutu-Upper Essequibo, Guyana

Cicadellidae: Cicadellinae
Oragua yuracensis • Sharpshooter
Boca Colorado, Madre de Dios, Peru

Cicadellidae: Cicadellinae
Oragua elegantula • Elegant Sharpshooter
Along the Rio Negro, Amazonas, Brazil

Cicadellidae: Cicadellinae
Parathona cayennensis • Cayenne Sharpshooter
Playa Los Angeles, Magdalena, Colombia

Cicadellidae: Cicadellinae
Paromenia auroguttata • Sharpshooter
Boquete, Chiriquí, Panama

Cicadellidae: Cicadellinae
Scopogonalia sp. • Sharpshooter
Filadelfia, Boquerón, Paraguay

Cicadellidae: Cicadellinae
Probably Paromenia sp. • Sharpshooter
Berg en Dal, Brokopondo, Suriname

Cicadellidae: Cicadellinae
Sibovia cf. conferta • Sharpshooter
El Hato, Sacatepéquez, Guatemala

Cicadellidae: Cicadellinae
Platygonia spatulata • Spatulate Sharpshooter
Valle de Antón, Coclé, Panama

Cicadellidae: Cicadellinae
Sibovia cf. festana • Sharpshooter
Tuis, Cartago, Costa Rica

Cicadellidae: Cicadellinae
Sibovia pileata • Sharpshooter
Tuis, Cartago, Costa Rica

Cicadellidae: Cicadellinae
Draeculacephala sp. • Sharpshooter
Tuis, Cartago, Costa Rica

Cicadellidae: Cicadellinae
Sibovia sagata • Sharpshooter
Colonia del Sacramento, Colonia, Uruguay

Cicadellidae: Cicadellinae
Stehlikiana halticula • Sharpshooter
Wayquecha Biological Resv., Cusco Department, Peru

Cicadellidae: Cicadellinae
Stephanolla rufoapicata • Sharpshooter
Valle de Antón, Coclé, Panama

Cicadellidae: Cicadellinae
Soosiulus sp. • Sharpshooter
Réserve Naturelle Trésor, French Guiana

Cicadellidae: Cicadellinae
Soosiulus sp. • Sharpshooter
Réserve Naturelle Trésor, French Guiana

Cicadellidae: Cicadellinae
Tylozygus geometricus • Sharpshooter
Selva Negra, Matagalpa, Nicaragua

Cicadellidae: Cicadellinae
Palingonalia bigutta • Sharpshooter
Reserva Particular do Cristalino, Mato Grosso, Brazil

Cicadellidae: Cicadellinae
Agrosoma sp. • Sharpshooter
Tuis, Cartago, Costa Rica

Cicadellidae: Cicadellinae
Tylozygus fasciatus • Sharpshooter
Pico Bonito National Park lodge, Atlántida, Honduras

Cicadellidae: Cicadellinae
Agrosoma sp. • Sharpshooter
Crooked Tree preserve, Belize district, Belize

Cicadellidae: Cicadellinae
Amblyscarta invenusta • Sharpshooter
Along the Rupununi River, Guyana

Cicadellidae: Cicadellinae
Cardioscarta sp. or Poeciloscarta sp. • Sharpshooter
Tuis, Cartago, Costa Rica

Cicadellidae: Cicadellinae
Apogonalia sanguinipes • Sharpshooter
Long Bay, Portland parish, Jamaica

Cicadellidae: Cicadellinae
Chlorogonalia coeruleovittata • Sharpshooter
Selva Negra, Matagalpa, Nicaragua

Cicadellidae: Cicadellinae
Cardioscarta quadrifasciata • Four-banded Sharpshooter
Along the Rupununi River, Guyana

Cicadellidae: Cicadellinae
Coronigoniella cf. maldonadoi • Sharpshooter
Along the Rupununi River, Guyana

Cicadellidae: Cicadellinae
Coronigoniella tiarae • Sharpshooter
Along the Rupununi River, Guyana

Cicadellidae: Cicadellinae
Erythrogonia cf. anduzei • Sharpshooter
Along the Rupununi River, Guyana

Cicadellidae: Cicadellinae
Dilobopterus sp. • Sharpshooter
Along the Rupununi River, Guyana

Cicadellidae: Cicadellinae
Erythrogonia cf. jucunda • Sharpshooter
Crooked Tree preserve, Belize district, Belize

Cicadellidae: Cicadellinae
Tettisama cf. quinquemaculata • Five-spot Sharpshooter
Reserva Particular do Cristalino, Mato Grosso, Brazil

Cicadellidae: Cicadellinae
Erythrogonia imitatricula • Sharpshooter
Wayquecha Biological Resv., Cusco department, Peru

Cicadellidae: Cicadellinae
Erythrogonia cf. laudata • Sharpshooter
Crooked Tree preserve, Belize district, Belize

Cicadellidae: Cicadellinae
Hortensia similis • Sharpshooter
Long Bay, Portland parish, Jamaica

Cicadellidae: Cicadellinae
Graphocephala alagarta • Sharpshooter
El Hato, Sacatepéquez, Guatemala

Cicadellidae: Cicadellinae
Ichthyobelus bellicosus • Bellicose Sharpshooter
Beside Lake Machuwasi, Madre de Dios., Peru

Cicadellidae: Cicadellinae
Graphocephala aurolineata • Sharpshooter
El Hato, Sacatepéquez, Guatemala

Cicadellidae: Cicadellinae
Inuyana cf. pendulosa • Sharpshooter
Boca Colorado, Madre de Dios, Peru

Cicadellidae: Cicadellinae
Jakrama sp. • Sharpshooter
Nanegalito, Tandayapa Valley, Ecuador

Cicadellidae: Cicadellinae
Macunolla ventralis • Sharpshooter
Valle de Antón, Coclé, Panama

Cicadellidae: Cicadellinae
Ladoffa cf. sannionis • Sharpshooter
Along the Rupununi River, Guyana

Cicadellidae: Cicadellinae: Proconiini
Abana cf. gigas • Sharpshooter
Tuis, Cartago, Costa Rica

Cicadellidae: Cicadellinae
Ladoffa cf. dependens • Sharpshooter
Leona Vicario, Quintana Roo, Mexico

Cicadellidae: Cicadellinae: Proconiini
Homalodisca cf. lucernaria • Sharpshooter
Playa Los Angeles, Magdalena, Colombia

Cicadellidae: Cicadellinae: Proconiini
Oncometopia aff. clarior • Sharpshooter
Selva Negra, Matagalpa, Nicaragua

Cicadellidae: Cicadellinae: Proconiini
Paracrocampsa sp. • Sharpshooter
Nanegalito, Tandayapa Valley, Ecuador

Cicadellidae: Cicadellinae: Proconiini
Oncometopia (Similitopia) sp. • Sharpshooter
Monteverde, Puntarenas, Costa Rica

Cicadellidae: Cicadellinae: Proconiini
Paraulacizes panamensis • Panama Sharpshooter
Boquete, Chriquí, Panama

Cicadellidae: Cicadellinae: Proconiini
Oncometopia sp. • Sharpshooter
Tuis, Cartago, Costa Rica

Cicadellidae: Cicadellinae: Proconiini
Raphirhinus phosphoreus • Needle-nosed Sharpshooter
Berg en Dal, Brokopondo, Suriname

Cicadellidae: Deltocephalinae
Bandaromimus sp. • Deltocephaline Leafhopper
Berg en Dal, Brokopondo, Suriname

Cicadellidae: Deltocephalinae
Osbornellus sp. • Deltocephaline Leafhopper
Selva Negra, Matagalpa, Nicaragua

Cicadellidae: Deltocephalinae
Taperinha sp. • Deltocephaline Leafhopper
Reserva Particular do Cristalino, Mato Grosso, Brazil

Cicadellidae: Deltocephalinae
Planicephalus flavicosta • Pale-margined Leafhopper
Crooked Tree preserve, Belize district, Belize

Cicadellidae: Deltocephalinae
Osbornellus sp. • Deltocephaline Leafhopper
Boquete, Chiriquí, Panama

Cicadellidae: Deltocephalinae
Stirellus picinus • Deltocephaline Leafhopper
Filadelfia, Boquerón, Paraguay

Cicadellidae: Deltocephalinae
Scaphytopius sp. • Deltocephaline Leafhopper
Suchitoto, Cuscatlán, El Salvador

Cicadellidae: Neocoelidiinae
Chinaia sp. • Neocoelidiine Leafhopper
Gamboa, Colón, Panama

Cicadellidae: Deltocephalinae
Tropicanus sp. • Deltocephaline Leafhopper
Mountain Pine Ridge Forest Reserve, Belize

Cicadellidae: Typhlocybinae
Protalebrella brasiliensis • Leafhopper
Crooked Tree preserve, Belize district, Belize

Cicadellidae: Neocoelidiinae
Neocoelidiinae • Neocoelidiine Leafhopper
Nanegalito, Tandayapa Valley, Ecuador

Membracidae: Centrotinae
Ischnocentrus niger • Treehopper
Tuis, Cartago, Costa Rica

Membracidae: Nicomiinae

Tolania sp. • Hornless Treehopper

Along the Rio Negro, Amazonas, Brazil

Membracidae: Membracinae

Enchenopa concolor • Treehopper

Montagne de Kaw, Roura, French Guiana

Membracidae: Membracinae

Aconophora laminate • Treehopper nymph

El Hato, Sacatepéquez, Guatemala

Membracidae: Membracinae

Enchenopa concolor • Treehopper

Novo Airão, Amazonas, Brazil

Membracidae: Membracinae

Campylenchia sp. • Treehopper

Tuis, Cartago, Costa Rica

Membracidae: Membracinae

Potnia brevicornis • Short-horned Treehopper

Monteverde, Puntarenas, Costa Rica

Membracidae: Smiliinae
Antianthe expansa • Keeled Treehopper
Selva Negra, Matagalpa, Nicaragua

Membracidae: Smiliinae
Horiola sp. • Treehopper
Montagne de Kaw, Roura, French Guiana

Membracidae: Smiliinae
Cyphonia sp. • Smiliine Treehopper
Along Arroyo de los Viboras, Colonia, Uruguay

Membracidae: Smiliinae
Neotynelia sp. • Treehoppers with ants
Leona Vicario, Quintana Roo, Mexico

Membracidae: Smiliinae
Cyphonia clavata • Clubbed Treehopper
Playa Los Angeles, Magdalena, Colombia

Membracidae: Smiliinae
Polyglypta sp. • Engraved Treehopper
Boquete, Chiriquí, Panama

Membracidae: Smiliinae
Polyglypta sp. • Engraved Treehopper
El Hato, Sacatepéquez, Guatemala

Membracidae: Stegaspidinae
Lycoderes sp. or Lycoderides sp. • Treehopper
Valle de Antón, Coclé, Panama

Membracidae: Smiliinae
Vestistilus sp. • Treehopper
Selva Negra, Matagalpa, Nicaragua

Membracidae: Stegaspidinae
≈ Stegaspis sp. • Stegaspidine Treehopper
Berg en Dal, Brokopondo, Suriname

Membracidae: Membracinae
Cladonota gonzaloi • Membracine Treehopper
El Hato, Sacatepéquez, Guatemala

Membracidae: Stegaspidinae
Stylocentrus championi • Champion's Stylocentrus
Tuis, Cartago, Costa Rica

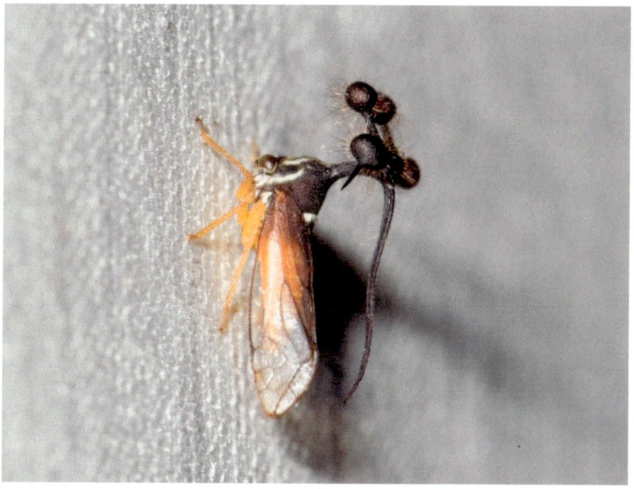

Membracidae: Stegaspidinae
Bocydium globulare • Globe-bearing Treehopper
Montagne de Kaw, Roura, French Guiana

Membracidae: Darninae
Darnis cf. lateralis • Darnine Treehopper
Along the Rupununi River, Guyana

Membracidae: Membracinae
Aconophorini • Treehopper
Altos del Maria, Coclé, Panama

Membracidae: Membracinae
Alchisme cf. grossa • Treehopper
Nanegalito, Tandayapa Valley, Ecuador

Membracidae: Membracinae
Membracis foliatafasciata • Treehopper
Novo Airão, Amazonas, Brazil

Membracidae: Membracinae
Membracis trimaculata • Three-spotted Membracis
Parque Nacional Tayrona, Magdalena, Colombia

Membracidae: Membracinae
Membracis sp. • Treehopper
Reserva Particular do Cristalino, Mato Grosso, Brazil

Cicadidae: Cicadettinae
Selymbria cf. pluvialis • Annual Cicada
Gamboa, Colón, Panama

Aetalionidae: Aetalioninae
Aetalion cf. reticulatum • Aetalionid Treehopper
Novo Airão, Amazonas, Brazil

Cicadidae: Cicadettinae
Taphura egeri • Annual Cicada
Rewa village, Upper Takutu-Upper Essequibo, Guyana

Cicadidae: Cicadettinae
Carineta sp. • Annual Cicada
Montagne de Kaw, Roura, French Guiana

Cicadidae: Cicadettinae
Taphura sp. • Annual Cicada
Montagne de Kaw, Roura, French Guiana

Cicadidae: Cicadinae
Fidicina mannifera • Annual Cicada
Rewa village, Upper Takutu-Upper Essequibo, Guyana

Cicadidae: Cicadinae
Fidicinini or Cryptotympanini • Annual Cicada
Pico Bonito National Park, Atlántida, Honduras

Cicadidae: Cicadinae
Fidicinoides pseudethelae • Annual Cicada
Rewa village, Upper Takutu-Upper Essequibo, Guyana

Aphididae
Aphididae • Aphid
Paracas, Ica, Peru

Cicadidae: Cicadinae
Fidicinini • Annual Cicada
Berg en Dal, Brokopondo, Suriname

Aphididae
Aphididae • Aphids with ant
Novo Airão, Amazonas, Brazil

Monophlebidae

Crypticerya sp. aff. abrahami • Scale insects with ants

Along the Rewa River, Guyana

Psyllidae: Psyllinae

Psyllinae • Plant Louse

Colonia del Sacramento, Colonia, Uruguay

Psyllidae

Psyllidae • Plant Louse

El Hato, Sacatepéquez, Guatemala

7

ORDERS MEGALOPTERA AND NEUROPTERA

Dobsonflies, Lacewings, Owlflies,

Mantisflies, and Antlions

Lacewings, dobsonflies, and their kin make up two relatively small orders, Megaloptera with about 300 described species worldwide, and Neuroptera with about 5000. Most are medium-sized to exceptionally large insects with wings that have a net-like pattern of veins. Both Megaloptera and Neuroptera have compound eyes that are separated; some have ocelli while others do not.

Megaloptera have long antennae, a short prothorax, and hyaline (colorless and transparent) wings. Males have a one-segmented cercus while females lack cerci. The order contains only two families, the alderflies (Sialidae) and the dobsonflies and fishflies (Corydalidae). Sialidae has a limited distribution in the Neotropics while Corydalidae are found widely in the region.

Most Neuroptera have long antennae that are thread-like or clubbed. Antlions have shorter antennae than the other families. Often various body surfaces are hairy in this order. Mouthparts are mandibulate. Neuroptera lack ovipositors and cerci. Among the families in Neuroptera are green lacewings (Chrysopidae), mantisflies (Mantispidae), owlflies (Ascalaphidae), and antlions (Myrmeleontidae).

Food. Adult Megaloptera are short-lived and most do not feed, or else consume only a little nectar. Larval food is discussed below.

In Neuroptera, owlflies are strong fliers and catch prey on the wing. Mantisflies share some anatomical features with the not-closely-related order Mantodea. Their raptorial forelegs and swiveling heads mean that they are fearsome predators of flies and other insects. Green lacewings are predatory and some feed on nectar and pollen. The short-lived adult antlions consume nectar and pollen; some species are predaceous as adults.

Reproduction. Megaloptera and Neuroptera undergo complete metamorphosis. The ancestral Megaloptera were the first insects to evolve holometabolism.

Details of courtship and mating are unknown for many Megaloptera and Neuroptera. Male dobsonflies in genus *Corydalus* first attempt to dislodge any rival males that may be perched near the female, using their enormous mandibles to flip competitors off their perch. The male then approaches the female from the side, touching her with his antennae. She responds with an apparent warning, opening her mandibles (which are much smaller than the male's) and moving her head from side to side. If she does not actually attack, the male lays his mandibles across the female, and soon moves into mating position.

In some dobsonfly taxa the male produces a sperm packet, part of which is edible. The packet is attached to the female's genital structures and then sperm transfer takes place. Later the female moves her abdomen close to her head and consumes the edible part, a nuptial gift.

Some green lacewings find partners for mating using vibrations. They vibrate their abdomens to make a certain pattern of pulses (unique to each species), and the abdominal vibrations cause the legs to vibrate. In turn, the vibrations of the legs are passed on to the leaf or other substrate and are received by the legs of the potential partner. By using these vibrations instead of audible sound, the lacewings avoid having bats or other predators listen in and discover their location. Lacewing females deposit a drop of glue on a leaf, then rapidly lay an egg on it and move their abdomen upward. The result is an egg that is elevated above the substrate, supported by a very slender stalk of dried glue. The stalk makes it difficult for ants or other predators to find or reach the eggs. The stalk is sometimes coated with a substance repellant to ants.

Development. Larvae in Neuroptera are prognathous (with forward-directed heads) and have mandibulate mouthparts that are modified for piercing-sucking. The gut does not connect to the outside of the larva's body and no solid waste is expelled until after the molt to adult stage.

Neuroptera have varied life histories. Antlion larvae prey upon ants, with some species digging conical pits for trapping their prey. Lacewing larvae feed most often on aphids and other Sternorrhyncha. Many mantisfly larvae feed on spider eggs and their development takes place inside spiders' egg sacs. Owlfly larvae are sit-and-wait predators that typically do their sitting and waiting from under leaf litter or other detritus. Most taxa in Neuroptera go through three larval instars.

The larvae of Megaloptera are aquatic, but they have both abdominal gills and spiracles and are sometimes seen on dry land near water. Larvae are prognathous and feed on various aquatic creatures including worms and larval insects. Pupation is terrestrial and takes place in chambers prepared in damp places such as in wet wood or under stones near a stream.

Larval development in Megaloptera takes one to four years as they move through up to eleven instars. The pupal stage, on the other hand, is quick, lasting as little as two weeks.

In both orders, pupae have freely moving legs and mandibles.

Dangers. Birds, reptiles, amphibians, and bats feed readily on these soft-bodied insects. Fish feed on larval Megaloptera. Insects also prey on the larvae and adults of the two orders.

Biting midges in subgenus *Pterobosca* attach themselves to the wings of neuropterans and suck the haemolymph. Parasitic wasps of at least two families attack the eggs of alderflies and dobsonflies.

Lacewings and other Neuroptera are attacked by entomopathogenic fungi in genera *Beauveria* and *Hirsutella.*

Defenses. Compared with most other orders, the insects in Neuroptera and Megaloptera are not excellent fliers when it comes to speed and agility. Thus, simply flying away from a predator often fails to take these insects to safety. In this respect, owlflies do better than other members of these two orders.

The high rates of predation on these insects is counterbalanced by impressive egg production. Just one dobsonfly may lay up to 3,000 eggs, for example, despite the short adult lifespan.

Green or brown coloration helps many species blend in with their substrate. Owlflies often position themselves on a branch in such a way that their abdomen strongly resembles a twig. Some mantisflies have evolved to look like paper wasps, predators that other predators tend to stay away from.

Most adult insects in these two orders have large, strong mandibles that are employed for defense. Larval mouthparts can also be formidable, as for example in the antlion family.

Larvae of green lacewings cover their backs with camouflage material, while some antlion larvae hunt with their bodies safely underground. Larval mantisflies in some taxa live and then pupate in the surprisingly safe environment of a spider egg sac.

Some lacewings secrete the substance skatole from their thorax. It has the smell of feces and repels most attackers. Lacewings in many taxa have a membrane for hearing built into their forewings, and they can hear the echolocation clicks of bats. They are then able to drop to the ground and avoid capture.

Relationship to humans. Most megalopterans and neuropterans have only a tenuous relationship with humans. The exception are green lacewing larvae (Chrysopidae), which farmers and gardeners value as predators of aphids and scale insects. These are sometimes released for natural pest control in commercial agriculture.

Figure 31 This larval lacewing has camouflaged its back with its own exuviae and possibly the drained bodies of its prey. Its killing mandibles are visible between the two leaf thorns. Belize.

Unfortunately, green lacewings are harmed by other biological control agents, particularly the entomopathogenic fungi released to control various true bugs and phytophagous beetles. Studies have shown that these fungi also attack beneficial lacewings, killing many and reducing the reproductive success of others.

The larvae of dobsonflies, called hellgrammites, are widely used as fishing bait.

Neotropical Specialties. Two genera of Megaloptera (*Protochauliodes* and *Archichauliodes*) are found only in Chile and Australia, suggesting an origin in Gondwanaland. An exception to the usually drab coloration in Megaloptera is provided by the Neotropical genus *Chloronia*, whose members are bright greenish-yellow. All three of these genera are in the dobsonfly family (Corydalidae).

Among the Neuroptera, the world's largest mantisfly, *Drepanicus gayi*, is found only in Chile. It resembles a green katydid. Some Neotropical antlion species have attractively patterned wings; these include *Glenurus spp.* and *Dimares spp.*

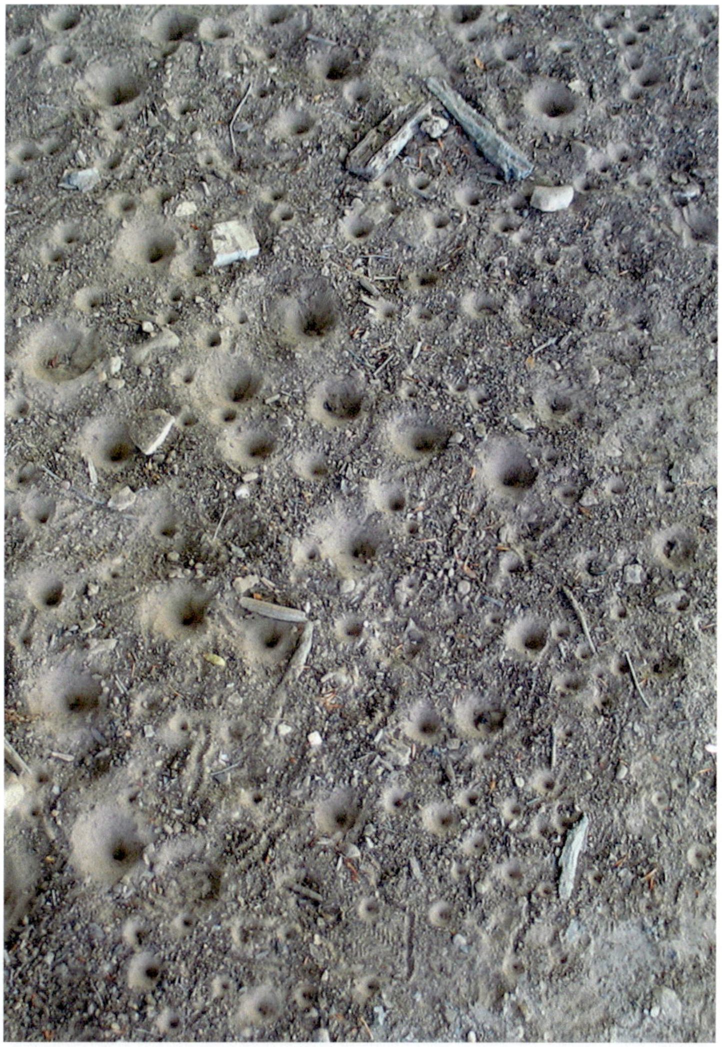

Figure 32 Conical pits dug by larval antlions (Myrmeleontinae). This site was protected from rain by an overhanging rock and had the ideal soil for pit traps. For these reasons, many antlions dug their pits here.

Chrysopidae: Chrysopinae
Ceraeochrysa cf. cubana • Cuban Lacewing
Mountain Pine Ridge Forest Reserve, Belize

Chrysopidae: Chrysopinae
Chrysoperla cf. externa • Green Lacewing
Colonia del Sacramento, Colonia, Uruguay

Chrysopidae: Chrysopinae
Plesiochrysa brasiliensis • Brazilian Lacewing
Montagne de Kaw, Roura, French Guiana

Chrysopidae: Chrysopinae
≈ Chrysopodes sp. • Green Lacewing
Playa los Angeles, Magdalena, Colombia

Corydalidae: Corydalinae
Corydalus sp. • Dobsonfly female
Tuis, Cartago, Costa Rica

Ascalaphidae: Ascalaphinae
Ascalaphinae • Split-eyed Owlfly
Colakreek, Para, Suriname

Ascalaphidae: Ascalaphinae
Ascalaphinae • Split-eyed Owlfly
Parque Nacional Pico Bonito, lodge, Atlántida, Honduras

Ascalaphidae: Haplogleniinae
Haplogleniinae • Owlfly
Tuis, Cartago, Costa Rica

Mantispidae: Mantispinae
Zeugomantispa compellens • Compelling Mantisfly
Tuis, Cartago, Costa Rica

Mantispidae: Mantispinae
≈ Dicromantispa sp. • Mantisfly
Montagne de Kaw, Roura, French Guiana

Mantispidae: Symphrasinae
Trichoscelia sp. • Mantisfly
Playa Los Angeles, Magdalena, Colombia

Myrmeleontidae: Myrmeleontinae
Myrmeleontinae • Antlion
Tuis, Cartago, Costa Rica

Mantispidae: Mantispinae
Climaciella brunnea • Brown Mantisfly (wasp mimic)
Parque Nacional Pico Bonito, lodge, Atlántida, Honduras

Corydalidae: Corydalinae
Corydalus sp. • Dobsonfly male
Tuis, Cartago, Costa Rica

Ascalaphidae
Ascalaphidae • Owlfly larva
Reserva Particular do Cristalino, Mato Grosso, Brazil

8

ORDER COLEOPTERA

BEETLES

No other insect order is as fully armored as Coleoptera. Beetles are covered with a tough, rigid, sclerotized material. Sclerotization is a process by which protein chains are cross-linked. Sclerotized substances are strong and durable; the keratin of horses' hooves is an example of a sclerotized material.

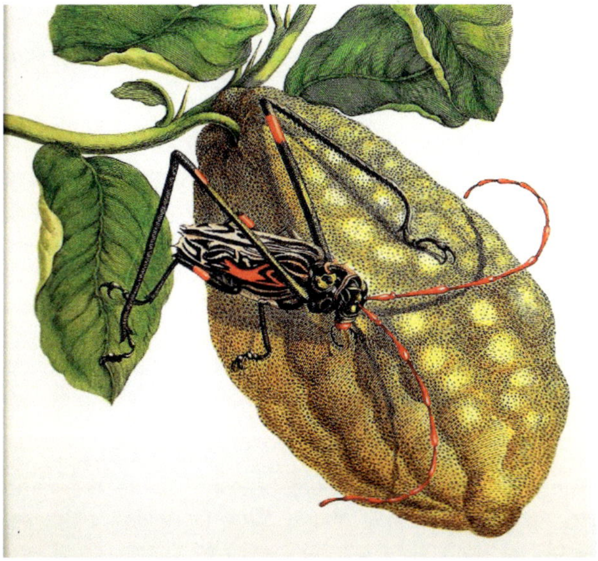

Figure 33 This is Maria Sibylla Merian's portrait of a male harlequin beetle *(Acrocinus longimanus)* in the long-horned beetle family. The long forelegs are used to "surround" a female to keep away rival males. Circa 1700, Suriname.

Beetles' extensive, sclerotized integument is difficult, often impossible, for predators to pierce. Taken together with other defenses such as chemical ones, and the abilities to fly, run, leap, and/or swim, the armored body sharply reduces other creatures' predation on beetles.

All insects have relatively tough exoskeletons, and the difference with beetles is one of degree. For example, beetles' body surfaces are much more sclerotized than those of mayflies, flies, or butterflies. There are also differences of degree *within* the order Coleoptera. For instance, fireflies are softer than other beetles and are sometimes described as "leathery." Fireflies' body coverings are not as hard and rigid as those of scarab beetles, but fireflies have stronger surfaces than aphids or termites.

The largest pieces of armor, the elytra, are also the most distinctive anatomical structure of beetles. The elytra are the front wings. In beetles these are not used to power flight, but to protect the hind wings and the abdomen. In some species elytra do provide lift from the angle of the leading edge that slices into the air. The elytra often curve downward to protect the beetle's sides and rear, and by covering the spiracles they can help prevent water loss in beetles living in arid regions.

The elytra meet in a straight line down the beetle's back, a key trait in identifying insects as beetles. The line where the elytra meet is called the elytral suture.

The pronotum is a dorsal plate of chitin that protects much of the beetle's thorax. Features of the pronotum are often useful in identifying a beetle to subfamily, genus, or species. Another structure often used in identifications is the scutellum, a small, often triangular chitinous structure located where the pronotum meets the elytral suture.

A third group of structures used often in beetle identification are the tarsi (the apical parts of the legs; "feet"). Literature on Coleoptera often discusses the tarsal formula, for example 5-5-4. The example describes a beetle that has 5 tarsomeres on each of the front two pairs of legs, and 4 tarsomeres on each hind leg. Sometimes a tiny or hidden tarsomere necessitates a formula like "appears as 4-4-4, but actually 5-5-5."

Beetles have mandibulate mouthparts used for chewing. Their antennae have 11 antennomeres in most groups, but a few taxa have more or fewer. Beetles have a pair of multifaceted compound eyes but almost all beetles lack ocelli.

Beetles are divided taxonomically into four suborders, two of which are quite small and not represented in the photos that follow. One fairly large suborder is Adephaga, made up of most aquatic beetle families as well as family Carabidae (ground beetles). The huge suborder Polyphaga is comprised of 149 families worldwide. Polyphaga includes about 90% of all beetle species.

Food. Beetles consume an astonishing variety of foodstuffs. Most Adephaga are predators. Most Polyphaga feed on various parts of plants, but a sizeable minority are predacious.

Among the foods consumed by beetles in Polyphaga are leaves, buds, stems, flowers, pollen, fruits, roots, galls, mushrooms, compost, dung, carrion, wood, fungal hyphae in wood, and stored products ranging from hams to flour to cigarettes. Predatory families in the suborder include rove beetles, soldier beetles, fireflies, and many others.

Some beetle species are omnivorous, while others specialize in only one foodstuff. The common names of taxa often reveal the preferred diet. Examples include carrion beetles (Silphidae), sap-feeding beetles (Nitidulidae), pleasing fungus beetles (Erotylidae), and dung beetles (Scarabaeinae and Aphodiinae).

Reproduction. Male and female beetles find each other in a variety of ways. Some species use visual cues, homing in on beetles with the same color, markings, and structural features. In other groups, females "call" for males by emitting an attractant pheromone. Fireflies, glowworms, and headlight beetles produce light to signal to potential mates.

In most groups male beetles mount and grip the female during mating. Males may remain attached to the female for some time, thus preventing other males from mating until the female lays the eggs that that male has fertilized.

Fireflies are a little different. The pair begins in the male-mounts-female position. At this stage, the female may reject the male. If the female chooses to continue, the couple soon moves to a tail-to-tail position, and in some cases may maintain that position for several hours.

Figure 34 Beetles mate in a male-mounts-female position. These are long-horned beetles, *Trichoxys abbreviatus*. Guatemala.

Development. Eggs are laid on, in, or near the larval food source. After hatching, the larvae feed voraciously and move though several instars (often five) as they grow. Since beetles are holometabolous, the final larval molt is to the pupal state. At the end of pupal development comes the molt to the adult stage.

Some larvae feed on the same foods utilized by adults, and in the same microhabitat. For instance, it is common to see adult and larval lady beetles side by side on a plant stem gobbling aphids. On the other hand, adults and larvae sometimes live markedly dissimilar lives. For example, some blister beetle hatchlings on flowers hitch a ride on an adult bee to the hive, then dine on larval bees and the food that had been prepared for the bee larvae. The adult blister beetles of these groups, on the other hand, feed on flowers and leaves.

Dangers. Mammals, birds, fish, frogs, turtles, lizards, spiders, and other insects are among the creatures that dine on adult and larval beetles.

Beetles are also harmed by parasites and parasitoids. Mites and pseudoscorpions are often seen riding on beetles. These sometimes feed on beetle haemolymph, but sometimes they are only using the beetle for transportation. Pseudoscorpions sometimes eat the mites that are attached to the beetle.

Beetles can become infested with entomopathogenic fungi. The spores fall onto the beetle and in conditions of high humidity they grow rapidly. Enzymatic activity allows the fungus to penetrate the external chitin and eventually reach the interior of the beetle's body. The beetle's death is caused by damage to the internal organs, or by toxins produced by the fungus. Affected beetles are often seen on leaves or tree trunks, covered with a carpet of pale fungus, perhaps with a stalked fruiting body rising above the insect's corpse.

Defenses. Two of the more obvious beetle defenses were mentioned briefly above. One is having a tough, sclerotized body that predators fail to crack, crush, or pierce. Some beetles whose elytra cover the sides as well as the back can sit down on the substrate and thus be almost flawlessly protected on all sides. The other defense mentioned above was fleeing by jumping, running, flying, or some swimming. One beetle family's common name refers to an escape method the family has perfected. These are the tumbling flower beetles (Mordellidae). When disturbed, these little beetles tuck in their legs and antennae and simply roll off the plant. They are almost impossible for the predator to find in the leaf litter below.

Some beetles are flat and can take refuge in rock crevices or under tree bark, where more normally proportioned predators cannot go. Some have spines that make them unpleasant or impossible to swallow.

As with other insect groups, so in the beetles many species use color and markings for defense. In some instances, the colors and markings allow them to blend in with their background. In other cases, bold colors and markings warn that the beetle is chemically protected. In still other cases the bold colors' warning is false but effective.

Figure 15 Two examples of beetle victims of entomopathogenic fungus. In both cases, the infections are far advanced, and the fungi have sent up fruiting bodies. Brazil.

/

Figure 36 This long-horned beetle *(Derobrachus sp.),* with its sharp thoracic spines, would not make a pleasant meal for a bird.

Chemical weapons among the beetles are varied both as to content and deployment. Sometimes beetles simply taste bad to predators because of substances within their bodies. Other beetles such as ladybirds and blister beetles can "reflex bleed" an irritating chemical from their legs if attacked.

Some beetles can even aim a caustic spray at an attacker. The most impressive of these are the bombardier beetles, a small group within the ground beetles. Bombardier beetles have two chemical storage tanks in their abdomen, one containing hydrogen peroxide and the other hydroquinone. If the beetle is attacked, both chemicals are released into an abdominal "combustion chamber" where they interact with an enzyme. An explosion of boiling hot quinones then exits the anus with a popping noise. The surprise, the heat, the noise, and the little cloud of noxious gases all help repel the attack. The beetle can even aim its anus to the front, sides, and back, and its aim is quite good. Some other beetles can expel chemicals to harm predators (certain darkling beetles, for example) but none as remarkably as the bombardiers.

Larval tortoise beetles deposit their feces on a forked structure above their back. A predator may only see feces, judging them to be unsuitable for food.

Relationship to humans. Beetles are of great economic importance. Some are pollinators. Others, such as ground beetles and lady beetles, eat creatures that attack crops, including aphids, caterpillars, snails, and slugs.

On the other hand, many phytophagous beetles feed on crops—often the leaves, but sometimes the roots, fruits, or other plant parts. Humans typically respond with insecticides, which can be effective but can also harm beneficial insects or enter the water supply.

Attempts by gardeners to purchase and put to work beneficial beetles, particularly lady beetles, usually meet with limited success. For one thing, the beetles may quickly leave the garden to feed elsewhere. Anyway, in a healthy ecosystem plants infested with aphids will quickly attract lady beetles and other aphidophagous insects.

Neotropical specialties. Several interesting groups of beetles are found only in the Neotropics (and sometimes in the extreme south of the United States too). These include the headlight beetles, which are click beetles (Elateridae) with two bioluminescent spots on the pronotum and one on the beetle's underside. The most common genus of headlight beetle is *Pyrophorus*. Entomologist Charles Hogue noted the flying beetles' "blue-green streak of light that dazzles the onlooker."

Other Neotropical specialties are particular species within widespread families. The ceiba borer *(Euchroma gigantea),* for example, is one of the world's largest buprestid beetles, with a length of 50 – 80 mm (about 2 - 3 inches). Its elytra have a wrinkled pattern and boast metallic green and red coloration. Tiger beetles are always a popular group among naturalists, but only in the Neotropics will you find *Pseudoxycheila tarsalis*, a beautiful deep blue species with two creamy round dots on the elytra. Tortoise beetles are found worldwide, but Neotropical species are often special favorites of naturalists, with some species having double-bullseye markings, and others having a pyramidal shape rather than the usual tortoise-like one.

One final point of interest among beetles in the Neotropics is that both the world's longest and shortest beetles are found there. The longest two beetles are the Hercules beetle (*Dynastes hercules*) and the titan beetle (*Titanus giganteus*). The former is a Scarab found in Central and South America whose length record is aided by an especially long horn. The Titan beetle of northern South America is a longhorn beetle that is hornless. Which beetle is longer depends on whether horns are included in the measurements. The world's record for the tiniest beetles goes to Neotropical members of the genus *Nanosella*, in the feather-winged beetles family (Ptiliidae). They measure up at just under 0.4 mm.

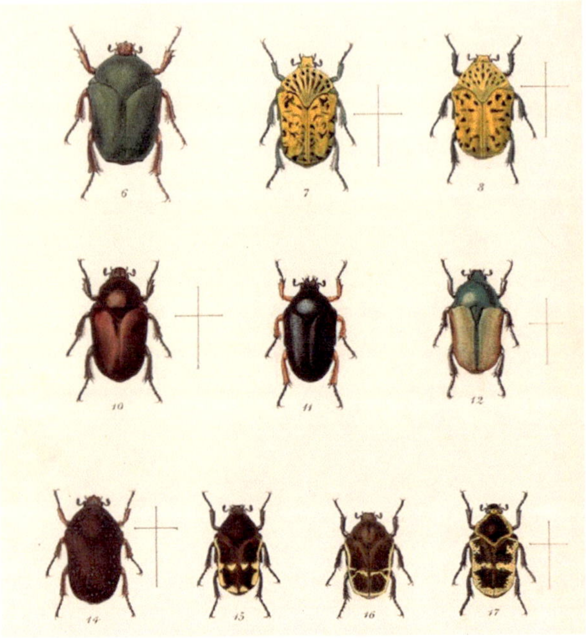

Figure 37 One of the beetle plates from the landmark work *Biologia Centrali-Americana* (published 1879 – 1915). This work comprised 63 volumes and depicted more than 18,000 plants and animals found in Central America. The volume that included this plate was written by the great collector of Neotropical beetles and butterflies, Henry Walter Bates.

Carabidae: Cicindelinae
Brasiella sp. • Brasiella Tiger Beetle
Lagunas de Aruza, Darién, Panama

Carabidae: Cicindelinae
Pseudoxycheila tarsalis • Blue Tiger Beetle
Tuis, Cartago, Costa Rica

Carabidae: Cicindelinae
Cicindela carthagena • Cartagena Tiger Beetle
Juan Hombrón, Coclé, Panama

Carabidae: Cicindelinae
Tetracha cf. sobrina • Big-headed Tiger Beetle
Rewa village, Upper Takutu-Upper Essequibo, Guyana

Carabidae: Cicindelinae
Cicindela trifasciata • S-banded Tiger Beetle
Laguna de Oviedo, Pedernales, Dominican Republic

Carabidae: Cicindelinae
Cylindera apiata • Tiger Beetle
Along the Río Uruguay, Colonia, Uruguay

Carabidae: Cicindelinae
Odontocheila sp. • Tiger Beetle
Réserve Naturelle Trésor, Roura, French Guiana

Carabidae: Cicindelinae
Odontocheila cayennensis • Cayenne Tiger Beetle
Berg en Dal, Brokopondo, Suriname

Carabidae: Cicindelinae
Odontocheila sp. • Tiger Beetle
Berg en Dal, Brokopondo, Suriname

Carabidae: Cicindelinae
Pentacomia sp. • Tiger Beetle
Sentier de Bagne des Annamites, French Guiana

Carabidae: Cicindelinae
Odontocheila cf. atriceps • Black-footed Tiger Beetle
Berg en Dal, Brokopondo, Suriname

Carabidae: Cicindelinae
Ctenostoma succinctum • Tiger Beetle
Montagne de Kaw, Roura, French Guiana

Carabidae: Scaritinae
≈ Scarites sp. • Large-headed Ground Beetle
Rewa village, Upper Takutu-Upper Essequibo, Guyana

Carabidae: Harpalinae
Notiobia cupreola • Coppery Ground Beetle
Tuis, Cartago, Costa Rica

Carabidae: Lebiinae
Pentagonica flavipes • Yellow-legged Ground Beetle
Presqu'île de la Caravelle, Martinique

Carabidae: Harpalinae
Notiobia chalcites • Ground Beetle
Filadelfia, Boquerón, Paraguay

Carabidae: Harpalinae
Calophaena nevermanni • Long-necked Ground Beetle
Tuis, Cartago, Costa Rica

Carabidae: Harpalinae
Selenophorus sp. • Harpaline Ground Beetle
Rewa village, Upper Takutu-Upper Essequibo, Guyana

Carabidae: Harpalinae
Selenophorus sp. • Harpaline Ground Beetle
Rewa village, Upper Takutu-Upper Essequibo, Guyana

Gyrinidae: Gyrininae
Dineutus longimanus • Whirligig Beetle
Bosque Estatal Río Abajo, Puerto Rico

Carabidae: Harpalinae
Agra sp. • Harpaline Ground Beetle
Berg en Dal, Brokopondo, Suriname

Hydrophilidae: Hydrophilinae
Berosus sp. • Water Scavenger Beetle
Rewa village, Upper Takutu-Upper Essequibo, Guyana

Dytiscidae: Copelatinae
Copelatus sp. • Predaceous Diving Beetle
Rewa village, Upper Takutu-Upper Essequibo, Guyana

Hydrophilidae: Hydropilinae
Hydrophilus cf. ensifer • Water Scavenger Beetle
Filadelfia, Boquerón, Paraguay

Silphidae: Silphinae
Oxelytrum discicolle • Carrion Beetle
Tuis, Cartago, Costa Rica

Passalidae: Passalinae
≈ Heliscus sp. • Patent-leather Beetle
Tuis, Cartago, Costa Rica

Histeridae: Tribalinae
Plagiogramma sp. • Clown Beetle
Along the Rewa River, Guyana

Staphylinidae: Paederinae
Paederus sp. • Paederine Rove Beetle
Selva Negra, Matagalpa, Nicaragua

Passalidae: Passalinae
Passalus cf. punctiger • Patent-leather Beetle
Along the Rewa River, Guyana

Staphylinidae: Paederinae
Paederus sp. • Paederine Rove Beetle
Tuis, Cartago, Costa Rica

Staphylinidae: Staphylininae
Leistotrophus versicolor • Versicolor Rove Beetle
Selva Negra, Matagalpa, Nicaragua

Scarabaeidae: Aphodiinae
≈ Ataenius sp. • Aphodiine Dung Beetle
Rewa village, Upper Takutu-Upper Essequibo, Guyana

Staphylinidae: Staphylininae
Xantholinini • Rove Beetle
Tuis, Cartago, Costa Rica

Scarabaeidae: Aphodiinae
Nialaphodius nigritus • Aphodiine Dung Beetle
Long Bay, Portland parish, Jamaica

Pselaphidae
Pselaphidae • Short-winged Mold Beetle
Rewa village, Upper Takutu-Upper Essequibo, Guyana

Scarabaeidae: Melolonthinae
Ceraspis sp. • Chafer
Paracas, Ica, Peru

Scarabaeidae: Melolonthinae
Hoplia sp. • Chafer
Santa Elena, Puntarenas, Costa Rica

Scarabaeidae: Scarabaeinae
Canthon cf. triangularis • Tumblebug
Sentier de Bagne des Annamites, French Guyana

Scarabaeidae: Scarabaeinae
Hansreia affinis • Tumblebug
Along the Rewa River, Guyana

Scarabaeidae: Scarabaeinae
Canthon sp. • Tumblebug
Along the Rupununi River, Guyana

Scarabaeidae: Scarabaeinae
Canthon septemmaculatus • Seven-spotted Tumblebug
Montagne de Kaw, Roura, French Guiana

Scarabaeidae: Scarabaeinae
Canthidium nr. deyrollei • Tumblebug
Réserve Naturelle Trésor, Roura, French Guiana

Scarabaeidae: Scarabaeinae
Canthidium sp. • Tumblebug
Along the Rewa River, Guyana

Scarabaeidae: Scarabaeinae
Oxysternon festivum • Dung Pellet Scarab
Along the Rewa River, Guyana

Scarabaeidae: Scarabaeinae
Deltochilum septemstriatum or D. guyane • Tumblebug
Sentier de Bagne des Annamites, French Guiana

Scarabaeidae: Scarabaeinae
Phanaeus cf. pyrois. • Horned Dung Scarab
Valle de Antón, Coclé, Panama

Scarabaeidae: Scarabaeinae
Ontherus cf. pseudodidymus • Dung-loving Scarab
Monteverde, Puntarenas, Costa Rica

Scarabaeidae: Melolonthinae
Macrodactylus sp. • Chafer
Nanegalito, Pichincha, Ecuador

Scarabaeidae: Melolonthinae
Macrodactylus sp. • Chafer
Nanegalito, Pichincha, Ecuador

Scarabaeidae: Melolonthinae
Isonychus sp. • Chafer
Tuis, Cartago, Costa Rica

Scarabaeidae: Melolonthinae
Macrodactylus costulatus • Chafer
Monteverde, Puntarenas, Costa Rica

Scarabaeidae: Melolonthinae
Isonychus ocellatus • Ocellated Chafer
Tuis, Cartago, Costa Rica

Scarabaeidae: Melolonthinae
Isonychus sp. • Chafer
Rewa village, Upper Takutu-Upper Essequibo, Guyana

Scarabaeidae: Melolonthinae
Barybas sp. • May Beetle
Tuis, Cartago, Costa Rica

Scarabaeidae: Cetoniinae
Marmarina sp. • Fruit Chafer
Puerto Madryn, Chubut, Argentina

Scarabaeidae: Rutelinae
Anomala sp. • Shining Leaf Chafer
Tuis, Cartago, Costa Rica

Scarabaeidae: Rutelinae
Anomala (cupricollis) sp. • Shining Leaf Chafer
Tuis, Cartago, Costa Rica

Scarabaeidae: Rutelinae
Anomala sp. • Shining Leaf Chafer
Nanegalito, Pichincha, Ecuador

Scarabaeidae: Rutelinae
Anomala sp. • Shining Leaf Chafer
Tuis, Cartago, Costa Rica

Scarabaeidae: Rutelinae
Anomala sp. • Shining Leaf Chafer
Crooked Tree Preserve, Belize district, Belize

Scarabaeidae: Rutelinae
Pelidnota (Chalcoplethis) sp. • Shining Leaf Chafer
Montagne de Kaw, Roura, French Guiana

Scarabaeidae: Rutelinae
Rutela laeta • Shining Leaf Chafer
Along the Rewa River, Guyana

Scarabaeidae: Rutelinae
Pelidnota belti • Shining Leaf Chafer
Tuis, Cartago, Costa Rica

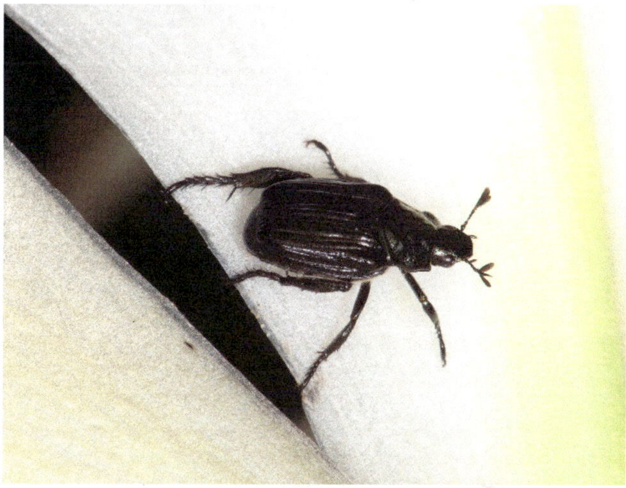

Scarabaeidae: Rutelinae
Strigoderma sp. • Shining Leaf Chafer
Monteverde, Puntarenas, Costa Rica

Scarabaeidae: Rutelinae
Pelidnota notata • Shining Leaf Chafer
Tuis, Cartago, Costa Rica

Scarabaeidae: Scarabaeinae
Ateuchus sp. • Dung-loving Scarab
Along the Rio Negro, Amazonas, Brazil

Scarabaeidae: Scarabaeinae
Bolbites onitoides • Dung-loving Scarab
Filadelfia, Boquerón, Paraguay

Scarabaeidae: Dynastinae
Cyclocephala brittoni • Dynastine Scarab
Sansón, Darién, Panama

Scarabaeidae: Dynastinae
Ancognatha vulgaris • Dynastine Scarab
Boquete, Chiriquí, Panama

Scarabaeidae: Dynastinae
Cyclocephala cf. discolor • Dynastine Scarab
Parque Nacional Pico Bonito, lodge, Honduras

Scarabaeidae: Dynastinae
Aspidolea fuliginea • Dynastine Scarab
Parque Nacional Pico Bonito, lodge, Honduras

Scarabaeidae: Dynastinae
Cyclocephala complanata • Dynastine Scarab
Parque Nacional Pico Bonito, lodge, Honduras

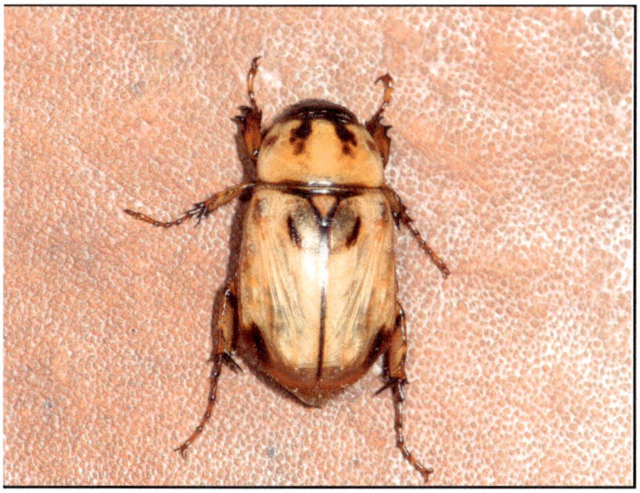

Scarabaeidae: Dynastinae
Cyclocephala amazona • Dynastine Scarab
Playa Los Angeles, Magdalena, Colombia

Scarabaeidae: Dynastinae
Cyclocephala sp. • Teneral Dynastine Scarab
Playa Los Angeles, Magdalena, Colombia

Scarabaeidae: Dynastinae
Cyclocephala mafaffa • Dynastine Scarab
Parque Nacional Pico Bonito, lodge, Honduras

Scarabaeidae: Dynastinae
Homophileurus cf. quadrituberculatus • Scarab
Montagne de Kaw, Roura, French Guiana

Scarabaeidae: Dynastinae
Cyclocephala tetrica • Dynastine Scarab
Long Bay, Portland parish, Jamaica

Scarabaeidae: Dynastinae
Megasoma elephas • Elephant Beetle ♀
El Llano, Panamá province, Panama

Scarabaeidae: Dynastinae
Strategus sp. • Ox Beetle
Novo Airão, Amazonas, Brazil

Buprestidae: Buprestinae
Euchroma gigantean • Giant Ceiba Borer
Along the Rupununi River, Guyana

Buprestidae: Buprestinae
Agrilaxia sp. • Buprestid Beetle
Colonia del Sacramento, Colonia, Uruguay

Buprestidae: Agrilinae
Agrilus sp. • Cane Borer
Crooked Tree Preserve, Belize district, Belize

Buprestidae: Buprestinae
Chrysobothris sp. • Sculptured Buprestid
Filadelfia, Boquerón, Paraguay

Buprestidae: Agrilinae
Leiopleura sp. • Buprestid Beetle
Colakreek, Para, Suriname

Limnichidae: Limnichinae
Limnichinae • Minute Marsh-loving Beetle
Rewa village, Upper Takutu-Upper Essequibo, Guyana

Limnichidae: Limnichinae
≈ Limnichites sp. • Minute Marsh-loving Beetle
Along Arroyo de las Víboras, Colonia, Uruguay

Ptilodactylidae: Ptilodactylinae
Ptilodactyla sp. • Toe-winged Beetle
Boquete, Chriquí, Panama

Elateridae: Agrypninae
Aeolus cf. panamensis • Click Beetle
Juan Hombrón, Coclé, Panama

Elateridae: Agrypninae
Pyrophorus sp. • Headlight Beetle
Arima Valley, Trinidad

Elateridae: Agrypninae
Agrypnus sp. • Click Beetle
Tuis, Cartago, Costa Rica

Elateridae: Agrypninae
Chalcolepidius sp. • Copper-scaled Click Beetle
Novo Airão, Amazonas, Brazil

Elateridae: Lissominae
Drapetes sp. • Click Beetle
Along the Rewa River, Guyana

Elateridae: Elaterinae
Dipropus sp. • Click Beetle
Along the Rewa River, Guyana

Elateridae: Semiotinae
Semiotus furcatus • Forked Semiotus
Reserva Particular do Cristalino, Mato Grosso, Brazil

Elateridae: Elaterinae
Glyphonyx sp. • Click Beetle
Boquete, Chiriquí, Panama

Elateridae: Semiotinae
Semiotus ligneus • Woody Semiotus
Montagne de Kaw, Roura, French Guiana

Elateridae: Denticollinae
≈ Atractosomus sp. • Click Beetle
Rewa village, Upper Takutu-Upper Essequibo, Guyana

Lycidae: Lycinae
Idiopteron sp. • Net-Winged Beetle
Berg en Dal, Brokopondo, Suriname

Lycidae: Platerodinae
≈ Plateros sp. • Net-winged Beetle
Boquete, Chiriquí, Panama

Lycidae: Lycinae
Calocladon sp. • Net-Winged Beetle
Leona Vicario, Quintana Roo, Mexico

Lycidae: Lycinae
Calopteron sp. • Net-Winged Beetle
Mountain Pine Ridge Forest Reserve, Belize

Lycidae: Lycinae
Mesopteron sp. • Net-winged Beetle
Selva Negra, Matagalpa, Nicaragua

Lycidae: Lycinae
Metapteron sp. • Net-winged Beetle
Along the Rupununi River, Guyana

Lampyridae: Lampyrinae
Photinus sp. • Firefly
Boquete, Chiriquí, Panama

Lampyridae: Lampyrinae
Photinini • Firefly
Along the Rupununi River, Guyana

Lampyridae: Lampyrinae
Photinus sp. • Firefly
Parque Nacional Pico Bonito, Atlántida, Honduras

Lampyridae: Lampyrinae
Photinini • Fireflies
Réserve Naturelle Trésor, Roura, French Guiana

Lampyridae: Lampyrinae
Photinus sp. • Firefly
Gamboa, Colón, Panama

Lampyridae: Lampyrinae
Photinus sp. • Firefly
Crooked Tree Preserve, Belize district, Belize

Lampyridae: Lampyrinae
Pyractonema sp. • Firefly
Torres del Paine National Park, Última Esperanza, Chile

Lampyridae: Lampyrinae
Photinus sp. • Firefly
Crooked Tree Preserve, Belize district, Belize

Lampyridae: Lampyrinae
Lucidota sp. • Fireflies
Boquete, Chiriquí, Panama

Lampyridae: Lampyrinae
Photinus sp. • Firefly
Tuis, Cartago, Costa Rica

Lampyridae: Lampyrinae
Lucidota sp. • Firefly
Selva Negra, Matagalpa, Nicaragua

Lampyridae: Lampyrinae

Lucidota sp. • Firefly

Montsinéry, Montsinéry-Tonnegrande, French Guiana

Lampyridae: Lampyrinae

Lamprocera sp. • Firefly

Parque Nacional Pico Bonito, Atlántida, Honduras

Lampyridae: Lampyrinae

Prob. Phaenolis sp. or Ophoelis sp. • Firefly

Réserve Naturelle Trésor, Roura, French Guiana

Lampyridae: Lampyrinae

Cratomorphus sp. • Firefly

Along the Rupununi River, Guyana

Lampyridae: Lampyrinae

Aspisoma sp. • Firefly

Filadelfia, Boquerón, Paraguay

Lampyridae: Lampyrinae

Tenaspis sp. • Firefly

Santo Domingo de los Colorados, Pichincha, Ecuador

Lampyridae: Photurinae
Bicellonycha sp. • Photurine Firefly
Boquete, Chiriquí, Panama

Cantharidae: Chauliognathinae
Chauliognathus segmentarius • Segmented Leatherwing
Parque Nacional Torres del Paine, Chile

Cantharidae: Chauliognathinae
Maronius brevipennis • Short-winged Soldier Beetle
Along the Rio Negro, Amazonas, Brazil

Cantharidae: Chauliognathinae
Chauliognathus sp. • Leatherwing
Boquete, Chiriquí, Panama

Cantharidae: Chauliognathinae
Lobetus cf. lemoulti • Short-winged Soldier Beetle
Sentier de Bagne des Annamites, French Guiana

Cantharidae: Chauliognathinae
Chauliognathus sp. • Leatherwing
Tuis, Cartago, Costa Rica

Cantharidae: Silinae
≈ Discodon sp. • Siline Soldier Beetle
Boquete, Chiriquí, Panama

Bostrichidae: Lyctinae
≈ Minthea sp. • Powder-post Beetle
Rewa village, Upper Takutu-Upper Essequibo, Guyana

Cantharidae: Silinae
Discodon sp. • Siline Soldier Beetle
Along the Rewa River, Guyana

Lymexylidae: Atractocerinae
Atractocerus sp. • Ship Timber Beetle
Montagne de Kaw, Roura, French Guiana

Cantharidae: Silinae
Silis barticana • Siline Soldier Beetle
Montagne de Kaw, Roura, French Guiana

Melyridae: Melyrinae
Astylus quadrilineatus • Soft-winged Flower Beetle
Nueva Palmira, Colonia, Uruguay

Nitidulidae: Cillaeinae
Macrostola cf. straminea • Sap-feeding Beetle
Montagne de Kaw, Roura, French Guiana

Nitidulidae: Nitidulinae
Stelidota sp. • Sap-feeding Beetle
Tuis, Cartago, Costa Rica

Nitidulidae: Epuraeinae
Epuraea cf. ocularis • Eyed Epuraea
Colonia del Sacramento, Colonia, Uruguay

Silvanidae: Brontinae
Telephanus sp. • Brontine Flat Bark Beetle
Boquete, Chiriquí, Panama

Nitidulidae: Nitidulinae
Prometopia sp. •Sap-feeding Beetle
Boquete, Chiriquí, Panama

Silvanidae: Silvaninae
Silvanus scuticollis • Flat Bark Beetle
Mountain Pine Ridge Forest Reserve, Belize

Erotylidae: Languriinae
Dasydactylus sellatus • Lizard Beetle
Crooked Tree Preserve, Belize district, Belize

Erotylidae: Erotylinae
Ischyrus quadripunctata • Four-spotted Ischyrus
Parque Nacional Tortuguero, Limón, Costa Rica

Erotylidae: Languriinae
Languriini • Lizard Beetle
Tuis, Cartago, Costa Rica

Erotylidae: Erotylinae
≈ Mycotretus sp. • Pleasing Fungus Beetle
Montsinéry, Montsinéry-Tonnegrande, French Guiana

Erotylidae: Erotylinae
Aegithus sp. • Pleasing Fungus Beetle
Tuis, Cartago, Costa Rica

Erotylidae: Erotylinae
Mycotretus sp. • Pleasing Fungus Beetle
Boquete, Chiriquí, Panama

Erotylidae: Erotylinae
Mycotretus maculatus • Pleasing Fungus Beetle
Novo Airão, Amazonas, Brazil

Erotylidae: Erotylinae
Iphiclus sp. • Pleasing Fungus Beetle
Along the Rewa River, Guyana

Erotylidae: Erotylinae
Mycotretus palmiphilus • Pleasing Fungus Beetle
Along the Rupununi River, Guyana

Erotylidae: Erotylinae
Iphiclus maculatus • Spotted Iphiclus
Along the Rewa River, Guyana

Erotylidae: Erotylinae
Mycotretus polyophtalmus • Many-eyed Fungus Beetle
Along the Rupununi River, Guyana

Erotylidae: Erotylinae
Iphiclus sp. • Pleasing Fungus Beetle
Along the Rio Negro, Amazonas, Brazil

Erotylidae: Erotylinae
Iphiclus sp. • Pleasing Fungus Beetle
Along the Rewa River, Guyana

Erotylidae • Erotylinae
Oligocorynus sp. • Pleasing Fungus Beetle
Tuis, Cartago, Costa Rica

Erotylidae: Erotylinae
Iphiclus sp. • Pleasing Fungus Beetle
Selva Negra, Matagalpa, Nicaragua

Erotylidae • Erotylinae
Oligocorynus zebra • Zebra Pleasing Fungus Beetle
Altos del Maria, Coclé, Panama

Erotylidae • Erotylinae
Iphiclus perspicillatus • Pleasing Fungus Beetle
Tuis, Cartago, Costa Rica

Bothrideridae: Bothriderinae
Lithophorus sp. • Dry Bark Beetle
Along the Rio Negro, Amazonas, Brazil

Coccinellidae: Coccinellinae
Curinus coeruleus • Metallic Blue Lady Beetle
Colonia del Sacramento, Colonia, Uruguay

Coccinellidae: Coccinellinae
Coleomegilla quadrifasciata • Four-banded Ladybird
Colonia del Sacramento, Colonia, Uruguay

Coccinellidae: Coccinellinae
Cycloneda sanguinea • Ladybird Beetle
Selva Negra, Matagalpa, Nicaragua

Coccinellidae: Coccinellinae
Eriopis connexa • Ladybird Beetle
Puerto Madryn, Chubut, Argentina

Coccinellidae: Coccinellinae
Olla v-nigrum • Ashy Gray Lady Beetle
Selva Negra, Matagalpa, Nicaragua

Coccinellidae: Coccinellinae
Neohalyzia perroudi • Ladybird Beetle
Boquete, Chiriquí, Panama

Coccinellidae: Coccinellinae
Psyllobora bicongregata • Fungus-eating Ladybird
Colonia del Sacramento. Colonia, Uruguay

Coccinellidae: Scymninae
Hyperaspidini • Scymnine Ladybird Beetle
Monteverde, Puntarenas, Costa Rica

Coccinellidae: Epilachninae
Epilachna tredecimnotata • Squash Lady Beetle
Playa Los Angeles, Magdalena, Colombia

Endomychidae: Lycoperdininae
Corynomalus archetypes • Handsome Fungus Beetle
Tuis, Cartago, Costa Rica

Coccinellidae: Scymninae
Cyrea arrowi • Scymnine Ladybird Beetle
Colonia del Sacramento, Colonia, Uruguay

Endomychidae: Stenotarsinae
Stenotarsus sp. • Handsome Fungus Beetle
Along the Rewa River, Guyana

Endomychidae: Epipocinae
Epopterus dives • Handsome Fungus Beetle
Reserva Particular do Cristalino, Mato Grosso, Brazil

Tenebrionidae: Stenochiinae
Strongylium maculicolle • Darkling Beetle
Monteverde, Puntarenas, Costa Rica

Tenebrionidae: Alleculinae
Lobopoda sp. • Comb-clawed Beetle
Reserva Particular do Cristalino, Mato Grosso, Brazil

Tenebrionidae: Stenochiinae
Talanus sp. • Darkling Beetle
Long Bay, Portland parish, Jamaica

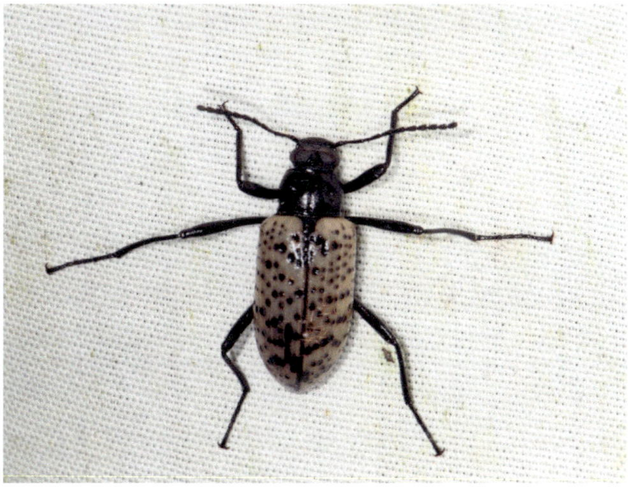
Tenebrionidae: Stenochiinae
Strongylium sallaei • Darkling Beetle
Parque Nacional Pico Bonito, lodge, Honduras

Tenebrionidae: Coelometopinae
Mylaris cf. gigas • Darkling Beetle
Novo Airão, Amazonas, Brazil

Tenebrionidae: Lagriinae
Prob. Goniadera sp. • Long-jointed Beetle
Along the Rio Negro, Amazonas state, Brazil

Tenebrionidae: Lagriinae
Statira sp. • Long-jointed Beetle
Playa Los Angeles, Magdalena, Colombia

Tenebrionidae: Lagriinae
Lagria villosa • Long-jointed Beetle
Filadelfia, Boquerón, Paraguay

Tenebrionidae: Diaperinae
Nilio sp. • False Lady Beetle
Montsinéry, Montsinéry-Tonnegrande, French Guiana

Tenebrionidae: Diaperinae
Nilio sp. • False Lady Beetle
Montsinéry, Montsinéry-Tonnegrande, French Guiana

Tenebrionidae: Diaperinae
Nilio sp. • False Lady Beetle
Reserva Particular do Cristalino, Mato Grosso, Brazil

Tenebrionidae: Pimeliinae
Scotobius pilularius • Darkling Beetle
Puerto Madryn, Chubut, Argentina

Tenebrionidae: Pimeliinae
Nyctelia granulata • Granulate Darkling Beetle
Isla Grande, Tierra del Fuego, Chile

Tenebrionidae: Phrenapatinae
Phrenapates bennetti • Darkling Beetle
Tuis, Cartago, Costa Rica

Mordellidae: Mordellinae
Mordellinae • Tumbling Flower Beetle
Playa Los Angeles, Magdalena, Colombia

Mordellidae: Mordellinae
Mordellinae • Tumbling Flower Beetle
Naranjo, Alajuela, Costa Rica

Oedemeridae: Oedemerinae
Oxycopis sp. • False Blister Beetle
Presqu'île de la Caravelle, Martinique

Oedemeridae: Oedemerinae
Diplectrus sp. • False Blister Beetle
Tuis, Cartago, Costa Rica

Cerambycidae: Prioninae
Orthomegas sp. • Prionine Longhorn
Berg en Dal, Brokopondo, Suriname

Cerambycidae: Prioninae
Callipogon cinnamomeus • Cinnamon Callipogon
Montagne de Kaw, Roura, French Guiana

Cerambycidae: Prioninae
Orthomegas marechali • Prionine Longhorn
Gamboa, Colón, Panama

Cerambycidae: Prioninae
Derobrachus sp. • Short-necked Prionine Longhorn
Tuis, Cartago, Costa Rica

Cerambycidae: Cerambycinae
Eburia cf. pedestris • Longhorned Beetle
Parque Nacional Pico Bonito, lodge, Honduras

Cerambycidae: Cerambycinae
Beraba cf. odettae • Long-horned Beetle
Playa Los Angeles, Magdalena, Colombia

Cerambycidae: Cerambycinae
Chlorida festiva • Festive Longhorn
Parque Nacional Pico Bonito, lodge, Honduras

Cerambycidae: Cerambycinae
Trichoxys abbreviatus • Longhorned Beetle
El Hato, Sacatepéquez, Guatemala

Cerambycidae: Cerambycinae
Chlorida cincta • Girdled Longhorn
Tuis, Cartago, Costa Rica

Cerambycidae: Cerambycinae
Tethlimmena aliena • Lycid-mimic Longhorn
Selva Negra, Matagalpa, Nicaragua

Cerambycidae: Cerambycinae
Elaphidiini • Long-horned Beetle
Long Bay, Portland parish, Jamaica

Cerambycidae: Cerambycinae
Elaphidion cf. jamaicensis • Long-horned Beetle
Long Bay, Portland parish, Jamaica

Cerambycidae: Cerambycinae
Trachyderes succinctus • Succinct Longhorned Beetle
Cabo Rojo, Pedernales, Dominican Republic

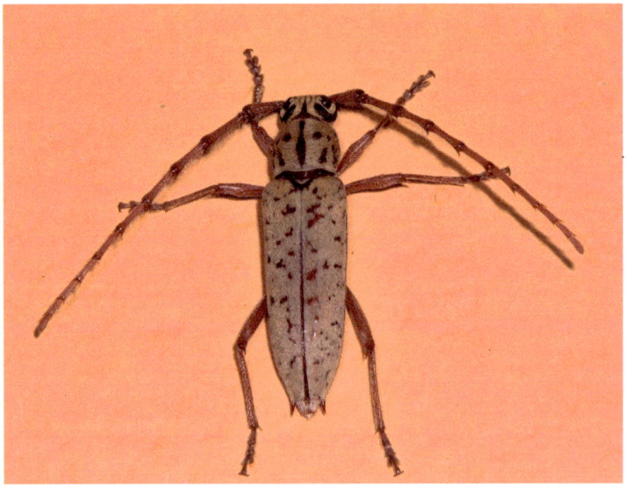

Cerambycidae: Cerambycinae
Elaphidion spinicorne • Long-horned Beetle
Long Bay, Portland parish, Jamaica

Cerambycidae: Lamiinae
Leptostylopsis sp. or Leptostylus sp. • Flat-faced LH
Long Bay, Portland parish, Jamaica

Cerambycidae: Cerambycinae
Gorybia sp. • Long-horned Beetle
Along the Rewa River, Guyana

Cerambycidae: Lamiinae
Onychocerus scorpio • Flat-faced Longhorn
Berg en Dal, Brokopondo, Suriname

Cerambycidae: Lamiinae
Oreodera sp. • Flat-faced Longhorn
Playa Los Angeles, Magdalena, Colombia

Cerambycidae: Lamiinae
Taeniotes sp. • Flat-faced Longhorn
Pantiacolla, Manú Road, Madre de Dios, Peru

Cerambycidae: Lamiinae
Taeniotes relucens • Flat-faced Longhorn
Reserva Particular do Cristalino, Mato Grosso, Brazil

Cerambycidae: Lamiinae
Taeniotes sp. • Flat-faced Longhorn
Parque Nactional Soberanía, Colón, Panama

Cerambycidae: Lamiinae
Taeniotes scalatus • Flat-faced Longhorn
Tuis, Cartago, Costa Rica

Cerambycidae: Lamiinae
Neoptychodes cretatus • Flat-faced Longhorn
Monteverde, Puntarenas, Costa Rica

Cerambycidae: Lamiinae
Onicideres cf. digna • Flat-faced Longhorn
Along the Teles Pires River, Mato Grosso, Brazil

Chrysomelidae: Criocerinae
Lema nr. elegantula • Criocerine Leaf Beetle
Tuis, Cartago, Costa Rica

Cerambycidae: Lamiinae
Xylotribus decorator • Flat-faced Longhorn
Along the Rupununi River, Guyana

Chrysomelidae: Criocerinae
Neolema nr. cordata • Criocerine Leaf Beetle
Gamboa, Colón, Panama

Disteniidae: Disteniinae
Paracometes cf. birai • Disteniid Longhorn
Tuis, Cartago, Costa Rica

Chrysomelidae: Criocerinae
Neolema dorsalis • Criocerine Leaf Beetle
Along the Rupununi River, Guyana

Chrysomelidae: Criocerinae
Neolema nr. sexpunctata • Criocerine Leaf Beetle
Parque Natural Metropolitano, Panamá, Panama

Chrysomelidae: Cassidinae
Chalepus nr. viduus • Hispine Leaf Beetle
Along the Rupununi River, Guyana

Chrysomelidae: Criocerinae
Neolema sp. • Criocerine Leaf Beetle
Filadelfia, Boquerón, Paraguay

Chrysomelidae: Cassidinae
Chalepus sp. • Hispine Leaf Beetle
Parque Nactional Tayrona, Magdalena, Colombia

Chrysomelidae: Cassidinae
Chalepispa ignorata • Hispine Leaf Beetle
Réserve Naturelle Trésor, Roura, French Guiana

Chrysomelidae: Cassidinae
Chalepus sp. • Hispine Leaf Beetle
Long Bay, Portland parish, Jamaica

Chrysomelidae: Cassidinae
Sceloenopla erudita or S. nevermanni • Hispine Beetle
Tuis, Cartago, Costa Rica

Chrysomelidae: Cassidinae
Agroiconota judaica • Judaic Tortoise Beetle
Montsinéry, Montsinéry-Tonnegrande, French Guiana

Chrysomelidae: Cassidinae
Sceloenopla nr. contraria • Hispine Beetle
Reserva Particular do Cristalino, Mato Grosso, Brazil

Chrysomelidae: Cassidinae
Aslamidium capense • Tortoise Beetle
Montsinéry, Montsinéry-Tonnegrande, French Guiana

Chrysomelidae: Cassidinae
Octotoma scabripennis • Rough-winged Hispine Beetle
El Hato, Sacatepéquez, Guatemala

Chrysomelidae: Cassidinae
Deloyala guttata • Tortoise Beetle
Tuis, Cartago, Costa Rica

Chrysomelidae: Cassidinae
Microctenochira diffinis • Tortoise Beetle
Berg en Dal, Brokopondo, Suriname

Chrysomelidae: Cassidinae
Microctenochira nr. semilunaris • Tortoise Beetle
Along the Rupununi River, Guyana

Chrysomelidae: Cassidinae
Microctenochira optata • Tortoise Beetle
Along the Rupununi River, Guyana

Chrysomelidae: Cassidinae
Microctenochira reticularis • Reticulate Tortoise Beetle
Berg en Dal, Brokopondo, Suriname

Chrysomelidae: Cassidinae
Microctenochira salebrata • Tortoise Beetle
Tuis, Cartago, Costa Rica

Chrysomelidae: Cassidinae
Helocassis crucipennis • Cross-winged Tortoise Beetle
Parque Nacional Tayrona, Magdalena, Colombia

Chrysomelidae: Cassidinae
Plagiometriona clavata • Clavate Tortoise Beetle
Selva Negra, Matagalpa, Nicaragua

Chrysomelidae: Cassidinae
Stolas discoides group • Tortoise Beetle
Berg en Dal, Brokopondo, Suriname

Chrysomelidae: Cassidinae
Cyrtonota lateralis • Lateral Tortoise Beetle
Berg en Dal, Brokopondo, Suriname

Chrysomelidae: Cassidinae
Stolas lebasii • Tortoise Beetle
Tuis, Cartago, Costa Rica

Chrysomelidae: Cassidinae
Stolas cf. brachiata • Tortoise Beetle
Tuis, Cartago, Costa Rica

Chrysomelidae: Cassidinae
Ischnocodia annulus • Ringed Tortoise Beetle
Along the Rupununi River, Guyana

Chrysomelidae: Cassidinae
Charidotis aurofasciata • Tortoise Beetle
Berg en Dal, Brokopondo, Suriname

Chrysomelidae: Cassidinae
Chelymorpha constellata • Starry Tortoise Beetle
Along the Rupununi River, Guyana

Chrysomelidae: Cassidinae
Charidotella sexpunctata • Golden Tortoise Beetle
Along the Rupununi River, Guyana

Chrysomelidae: Cassidinae
Chelymorpha marginata • Margined Tortoise Beetle
Berg en Dal, Brokopondo, Suriname

Chrysomelidae: Cassidinae
Charidotella sinuata • Sinuous Tortoise Beetle
Tuis, Cartago, Costa Rica

Chrysomelidae: Cassidinae
Imatidium thoracicum • Thoracic Tortoise Beetle
Gamboa, Colón, Panama

Chrysomelidae: Cassidinae
Parimatidium sp. • Tortoise Beetle
Along the Rio Negro, Amazonas, Brazil

Chrysomelidae: Galerucinae
Coelomera cf. cayennensis • Cayenne Leaf Beetle
Parque Nacional Tayrona, Magdalena, Colombia

Chrysomelidae: Cassidinae
Zatrephina lineata • Tortoise Beetle
Montsinéry, Montsinéry-Tonnegrande, French Guiana

Chrysomelidae: Galerucinae
Dircema cf. marginatum • Margined Galerucine Beetle
Along the Rupununi River, Guyana

Chrysomelidae: Galerucinae
Cerotoma salvini • Skeletonizing Leaf Beetle
Juan Hombron, Coclé, Panama

Chrysomelidae: Galerucinae
Dircema nigripenne • Black-winged Galerucine Beetle
Berg en Dal, Brokopondo, Suriname

Chrysomelidae: Galerucinae
Exora olivacea • Skeletonizing Leaf Beetle
Along the Rewa River, Guyana

Chrysomelidae: Galerucinae
Paranapiacaba dorsoplagiata • Skeletonizing Leaf Beetle
Boquete, Chiriquí, Panama

Chrysomelidae: Galerucinae
Exora encaustica • Caustic Leaf Beetle
Berg en Dal, Brokopondo, Suriname

Chrysomelidae: Galerucinae
Diabrotica biannularis • Two-ringed Diabrotica
Monteverde, Puntarenas, Costa Rica

Chrysomelidae: Galerucinae
Metacycla sp. • Skeletonizing Leaf Beetle
El Hato, Sacatepéquez, Guatemala

Chrysomelidae: Galerucinae
Diabrotica josephbalyi • Baly's Leaf Beetle
Tuis, Cartago, Costa Rica

Chrysomelidae: Galerucinae
Diabrotica adelpha • Skeletonizing Leaf Beetle
Crooked Tree Preserve, Belize district, Belize

Chrysomelidae: Galerucinae
Diabrotica fulvicornis • Skeletonizing Leaf Beetle
Tuis, Cartago, Costa Rica

Chrysomelidae: Galerucinae
Diabrotica balteata • Banded Cucumber Beetle
Tuis, Cartago, Costa Rica

Chrysomelidae: Galerucinae
Diabrotica signifera • Skeletonizing Leaf Beetle
Leona Vicario, Quintana Roo, Mexico

Chrysomelidae: Galerucinae
Diabrotica cf. circulata • Skeletonizing Leaf Beetle
Fortín de las Flores, Veracruz, Mexico

Chrysomelidae: Galerucinae
Diabrotica cf. sinuata • Skeletonizing Leaf Beetle
El Hato, Sacatepéquez, Guatemala

Chrysomelidae: Galerucinae
Diabrotica nr. boliviana • Skeletonizing Leaf Beetle
Along the Rio Negro, Amazonas, Brazil

Chrysomelidae: Galerucinae
≈ Walterianella sp. • Flea Beetle
Along the Rewa River, Guyana

Chrysomelidae: Galerucinae
Disonycha glabrata • Pigweed Flea Beetle
Metetí, Darién, Panama

Chrysomelidae: Galerucinae
Gynandrobrotica cf. jucunda • Flea Beetle
Playa Los Angeles, Magdalena, Colombia

Chrysomelidae: Galerucinae
≈ Walterianella sp. • Flea Beetle
Along the Rupununi River, Guyana

Chrysomelidae: Galerucinae
Kuschelina adjunctus • Flea Beetle
Isla del Sol, Lago Titicaca, Bolivia

Chrysomelidae: Galerucinae
Acanthonycha sp. • Flea Beetle
Tuis, Cartago, Costa Rica

Chrysomelidae: Galerucinae
Prob. Allochroma lunatum • Moon-spot Leaf Beetle
Montsinéry, Montsinéry-Tonnegrande, French Guiana

Chrysomelidae: Galerucinae
Omophoita aequinoctialis • Equinoctial Flea Beetle
Tuis, Cartago, Costa Rica

Chrysomelidae: Galerucinae
Systena sp. • Flea Beetle
Montsinéry, Montsinéry-Tonnegrande, French Guiana

Chrysomelidae: Galerucinae
Omophoita nr. cyanipennis • Flea Beetle
Long Bay, Portland parish, Jamaica

Chrysomelidae: Galerucinae
Paratriarius cf. adonis • Skeletonizing Leaf Beetle
Santo Domingo de los Colorados, Pichincha, Ecuador

Chrysomelidae: Galerucinae
Trichobrotica brasiliensis • Skeletonizing Leaf Beetle
Berg en Dal, Brokopondo, Suriname

Chrysomelidae: Galerucinae
Alagoasa fasciata • Banded Alagoasa
Maniche, département du Département Sud, Haiti

Chrysomelidae: Galerucinae
Asphaera discicollis • Flea Beetle
Berg en Dal, Brokopondo, Suriname

Chrysomelidae: Galerucinae
≈ Alagoasa sp. • Flea Beetle
Parque Nacional Tayrona, Magdalena, Colombia

Chrysomelidae: Galerucinae
Alagoasa sp. • Flea Beetle
Novo Airão, Amazonas, Brazil

Chrysomelidae: Galerucinae
Asphaera sp.or Alagoasa sp. • Flea Beetle
Montsinéry, Montsinéry-Tonnegrande, French Guiana

Chrysomelidae: Galerucinae
Capraita nr. nigrosignata • Flea Beetle
Along the Rio Negro, Amazonas, Brazil

Chrysomelidae: Cryptocephalinae
Pachybrachis sp. • Case-bearing Leaf Beetle
Montagne de Kaw, Roura, French Guiana

Chrysomelidae: Galerucinae
Capraita sp. • Flea Beetle
Pantiacolla, Manú Road, Madre de Dios, Peru

Chrysomelidae: Cryptocephalinae
Griburius triguttatus • Case-bearing Leaf Beetle
Berg en Dal, Brokopondo, Suriname

Chrysomelidae: Galerucinae
Monocesta depressa • Skeletonizing Leaf Beetle
Tuis, Cartago, Costa Rica

Chrysomelidae: Chrysomelinae
Platyphora luteipennis • Chrysomeline Leaf Beetle
Nanegalito, Pichincha, Ecuador

Chrysomelidae: Chrysomelinae
Platyphora sexmaculata • Chrysomeline Leaf Beetle
Tres Cruces, Parque Nacional Manú, Peru

Chrysomelidae: Chrysomelinae
Leptinotarsa undecimlineata • Eleven-lined Potato Beetle
Long Bay, Portland parish, Jamaica

Chrysomelidae: Chrysomelinae
Platyphora staudingeri • Chrysomeline Leaf Beetle
Salvador, Manú National Park, Peru

Chrysomelidae: Eumolpinae
Chalcophana cf. porcaticolor • Eumolpine Leaf Beetle
Tuis, Cartago, Costa Rica

Chrysomelidae: Chrysomelinae
Platyphora spectabilis • Chrysomeline Leaf Beetle
Tuis, Cartago, Costa Rica

Chrysomelidae: Eumolpinae
Lamprosphaerus sp. • Eumolpine Leaf Beetle
Along the Rupununi River, Guyana

Chrysomelidae: Eumolpinae
Eumolpini section Iphimeites • Eumolpine Leaf Beetle
Rewa village, Upper Takutu-Upper Essequibo, Guyana

Chrysomelidae: Eumolpinae
Sternurus sp. or Belta sp. • Eumolpine Leaf Beetle
Novo Airão, Amazonas, Brazil

Chrysomelidae: Eumolpinae
Colaspis sp. • Eumolpine Leaf Beetle
Along the Rupununi River, Guyana

Chrysomelidae: Eumolpinae
≈Metaxyonycha sp. • Eumolpine Leaf Beetle
Filadelfia, Boquerón, Paraguay

Chrysomelidae: Eumolpinae
Spintherophyta sp. • Eumolpine Leaf Beetle
Réserve Naturelle Trésor, Roura, French Guiana

Rhynchitidae: Rhynchitinae
Eugnamptus sp. • Tooth-nosed Snout Weevil
Boquete, Chiriquí, Panama

Brentidae: Brentinae
Brentus guatemalenus • Straight-snouted Weevil
Selva Negra, Matagalpa, Nicaragua

Brentidae: Brentinae
Raphirhynchus sp. • Straight-snouted Weevil
Reserva Particular do Cristalino, Mato Grosso, Brazil

Brentidae: Brentinae
Brentus cylindrus • Straight-snouted Weevil
Along the Rupununi River, Guyana

Curculionidae: Entiminae
Compsus sp. • Broad-nosed Weevil
Along the Rewa River, Guyana

Brentidae: Brentinae
Nemorhinus myrmecophagus • Straight-snouted Weevil
Rewa village, Upper Takutu-Upper Essequibo, Guyana

Curculionidae: Entiminae
Exophthalmus scalptus • Broad-nosed Weevil Monteverde,
Puntarenas, Costa Rica

Curculionidae: Entiminae
Naupactus sp. sensu lato • Broad-nosed Weevil Playa Los Angeles, Magdalena, Colombia

Curculionidae: Brachycerinae
Plectrophoroides sp. • Brachycerine Weevil
Along the Rio Negro, Amazonas, Brazil

Curculionidae: Entiminae
≈ Naupactus sp. • Broad-nosed Weevil
Along the Rupununi River, Guyana

Curculionidae: Cryptorhynchinae
Collabismus sp. • Hidden-snout Weevil
Berg en Dal, Brokopondo, Suriname

Curculionidae: Dryophthorinae
Rhinostomus barbirostris • Bearded Weevil
Playa Los Angeles, Magdalena, Colombia

Curculionidae: Dryophthorinae
Metamasius hemipterus • West Indian Sugarcane Weevil
Novo Airão, Amazonas, Brazil

Curculionidae: Dryophthorinae
Metamasius cf. dasyurus • Dryophthorine Weevil
Altos del Maria, Coclé, Panama

Curculionidae: Conoderinae
Hoplocopturus varipes • Conoderine Weevil
Selva Negra, Matagalpa, Nicaragua

Curculionidae: Dryophthorinae
Paramasius distortus • Palm Weevil
Montsinéry, Montsinéry-Tonnegrande, French Guiana

Curculionidae: Conoderinae
Piazurus sacer • Conoderine Weevil
Berg en Dal, Brokopondo, Suriname

Curculionidae: Conoderinae
Hoplocopturus cf. ochraceicollis • Conoderine Weevil
Altos del Maria, Coclé, Panama

Curculionidae: Conoderinae
Piazurus sp. • Conoderine Weevil
Tuis, Cartago, Costa Rica

Curculionidae: Conoderinae
Macrocopturus sp. • Weevil
Along the Rupununi River, Guyana

Curculionidae: Molytinae
Cholus lecidiosus • Molytine Weevil
Tuis, Cartago, Costa Rica

Curculionidae: Conoderinae
Cratosomus sp. • Conoderine Weevil
Along the Rupununi River, Guyana

Curculionidae: Molytinae
Cholus cinctus • Molytine Weevil
Altos del Maria, Coclé, Panama

Curculionidae: Cyclominae
Aegorhinus vitulus • El Caballito / Little Horse
Reserva Forestal Magallanes, Magallanes, Chile

Curculionidae: Molytinae
Heilipus sp. • Molytine Weevil
Nanegalito, Pichincha, Ecuador

Curculionidae: Molytinae
Heilipodus sp. • Molytine Weevil
Along the Rupununi River, Guyana

Curculionidae: Molytinae
Lixodes taeniatus • Molytine Weevil
Colonia del Sacramento, Colonia, Uruguay

Curculionidae: Molytinae
Hilipinus occultus • Molytine Weevil
Monteverde, Puntarenas, Costa Rica

Curculionidae: Molytinae
Marshallius rusticus • Molytine Weevil
Long Bay, Portland parish, Jamaica

Curculionidae: Molytinae
Homalinotus sp. • Molytine Weevil
Along the Rewa River, Guyana

Curculionidae: Ceratopodinae
Ceratopus sp. • Fig Weevil
Tuis, Cartago, Costa Rica

Curculionidae: Baridinae
Peridinetus coccineifrons • Red-browed Weevil
Tuis, Cartago, Costa Rica

Curculionidae: Platypodinae
Euplatypus parallelus • Pinhole Bark Borers
Rewa village, Upper Takutu-Upper Essequibo, Guyana

Curculionidae: Scolytinae
≈ Xyleborus sp. • Weevil
Tuis, Cartago, Costa Rica

9

ORDER DIPTERA

FLIES

Humans have noticed, discussed, and studied flies for millennia. Though small creatures, flies' impact on people is great. On the negative side of the ledger, some fly families include vectors of human diseases. Fly larvae in many groups are destroyers of crops and pests of farm animals. Flies annoy humans when they bite, and even when they perch repeatedly on people's bodies to lap up sweat. On the positive side, they are important pollinators (second only to Hymenoptera). Fly larvae often consume insect pests such as aphids. And without flies' cleaning and recycling services, planet earth would have an overabundance of putrescent carcasses, odiferous dung heaps, and mountains of rotting plants.

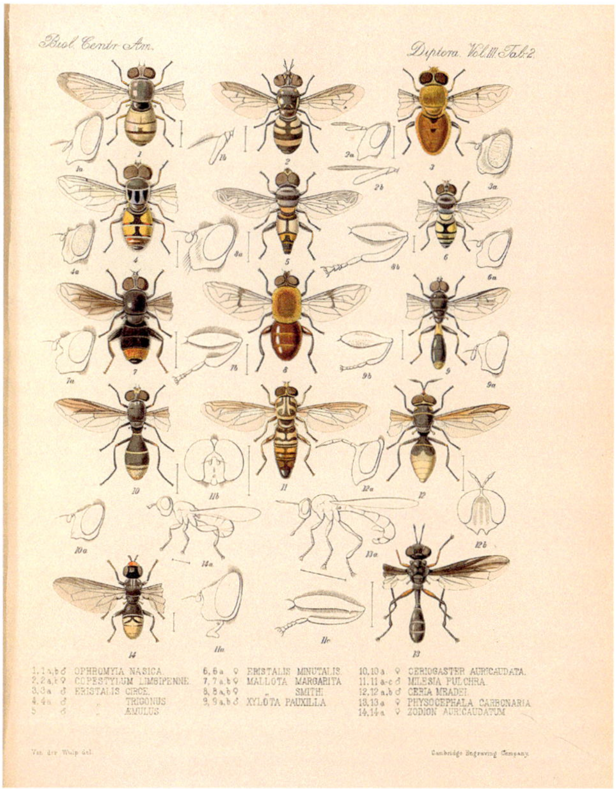

Figure 38 A plate from one of the three Diptera volumes of *Biologia Centrali-Americana,* which were published between 1886 and 1903.

While most people think wholly negative thoughts about flies, the bright colors, the spots and stripes, and the zigzags on the eyes of many species make this as beautiful and interesting a group as the dragonflies or beetles. Most fly species look quite different from the house fly we tend to think of as the quintessential dipteran. It is fun to observe these creatures' acrobatic flights and their ability to walk on the ceiling, to hover in one spot for long periods, or to fly off in hot pursuit of some tiny six-legged prey. Watching courtship, egg-laying, and larval feeding is an interesting pursuit, and can add to our understanding of poorly known species. The larvae of most of the described fly species are unknown to science.

Flies are remarkable for their sheer numbers and great diversity. In his 1980 entomology textbook Cedric Gillot wrote, "The more than 80,000 species of Diptera described to date probably represents about half of the world total." By 2012 Stephen Marshall was able to state that 160,000 fly species had been described, but that "we probably share the planet with between 400,000 and 800,000 fly species." Marshall estimated that as many as one-fifth of the animal species on earth are flies.

The name of the order, Diptera, is an ancient Greek construction that means "two wings." Almost all flies have two wings, and instead of a rear pair of wings they have two "halteres"— structures that look like a knobbed stick, and that play a role in the insect's equilibrium in flight. No flies have four fully developed wings, while several small groups are wingless.

Members of this order are relatively soft-bodied, and tend to have large compound eyes, three ocelli, and a head that is able to move in many directions. Characterisitcs of the antennae are important in fly identification. Nematocera (crane flies, midges, gnats, and mosquitoes) have long, slender antennae of nine or more antennomeres, while more recently evolved groups like robber flies, bee flies, and house flies have fewer (commonly three to six). Some of the more advanced groups feature an "arista" which is a hair-like structure attached to each antenna. Traits of the arista are also important in fly taxonomy.

Fly mouthparts are suited for liquid foods. In some groups the mouthparts are specialized for piercing-sucking; in others, for licking-lapping. Flies may regurgitate digestive fluids onto solid food to begin the process of liquefying it.

The wings of flies are often short when compared with those of other insect orders, but the flying abilities of dipterans are amazing. Their speed is remarkable, the number of wingbeats per second can be hard to believe, changes of direction are accomplished almost instantly, and flies have even been observed flying backwards.

Six bristly legs help some predacious species capture their prey in a basket-like configuration. In some groups the legs have ornaments that are used in sexual displays. All flies have five tarsomeres on each leg.

Figure 39 Flies in suborder Nematocera are typically slender and have long legs, though not always as long as those on this limoniid crane fly, *Trentepohlia sp.,* from French Guiana.

Food. Adult and larval flies consume a surprisingly wide variety of foods. Among these are dung, carrion, the blood of vertebrates, yeasts and other fungi, compost, leaves, fruits, sap, pollen, and nectar. Predatory flies attack snails, slugs, worms, and insects. A few groups (often wingless) are external parasites of birds or mammals, clinging to their host and taking repeated blood meals. Other groups, like the large family Tachinidae, are parasitoids that develop on or in caterpillars, true bugs, grasshoppers, and other insects.

Reproduction. Courtship rituals are fairly common in order Diptera. Some *Drosophila spp.* fruit flies engage in a courtship dance that is begun by the male and joined in by the female. The dance consists primarily of sideways steps with regular changes of direction. Some male flies display their wing patterns or odd leg structures to the female for her approval. Flies in family Empididae are called dance flies because the males do an airbone up-and-down dance in large swarms with others of their species. The swarm may appear like a cloud of smoke, forming a temporary landmark that females can easily spot.

Ovipositing occurs in a location where the larvae will be able to find their preferred foodstuff.

Figure 40 Typical mating pair of flies, showing the male-mounts-female posture found in Diptera. These are drosophilid fruit flies, probably *Hirtodrosophila sp.* Mexico.

Development. Fly larvae are either fully aquatic or else they develop in damp locations such as mud or dung. The larvae of most Nematocera are aquatic, and most move through four larval instars. In other groups the larvae are less often fully aquatic; in these non-Nematocera groups larvae pass through five to eight instars.

Fly larvae vary considerably in form. In Nematocera the larvae have a distict head capsule and their jaws move in a horizontal plane. Other groups have their heads less-developed and jaws that move in a vertical plane. Fly larvae are often called maggots, and they can seem to have few discernable body parts. Under magnification the anatomical structures are more recognizable.

One unusual trait possessed by many dipterans is mobility in the pupal stage. While the pupae in other orders are immobile or nearly so, many fly pupae are quite active. Mosquito pupae, for example, are aquatic and must regularly swim to the surface to replenish their oxygen supply. Robber flies (Asilidae) spend their pupal stage underground, but are able to climb many inches to the surface using pupal spines to push against the soil.

The most advanced flies, such as flesh flies and house flies, have a pupal stage that takes place within the last larval "skin," which serves as a protective puparium. Some primitive groups also have unusual pupal characteristics. Black flies (Simuliidae) have aquatic pupae that inflate with air at the end of the pupal period. The inflation causes the pupal case to split, and the adult black fly rises to the water's surface in a bubble of air, emerging dry and almost ready for the first flight.

Dangers for Diptera. Fly larvae, pupae, and adults are favorite foods of many birds, bats, amphibians, and reptiles. (Commercially sold turtle food consists of fly pupae.) Fish, spiders, and other insects also relish flies as a part of their diets.

Figure 11 Jumping spiders (Salticidae) are successful predators against soft-bodied insects like flies. Suriname.

Humans often respond to problems that flies cause by reaching for insecticides. DDT was widely used to control mosquitoes and other biting flies, causing not only a reduction in the number of flies but a serious reduction in the number of birds and other organisms. While safer insecticides are available today, they still can be harmful to humans and to the ecosystem, and need to be used wisely and with environmental monitoring.

Flies also fall victim to entomopathogenic fungi which leave them barely able to move, and eventually kills them.

Defenses. The primary defense of flies is their flying abilities. Many predators lack the flight skills that flies possess, and catch them only occasionally. Flies can often escape predators with their impressive flight speeds and ability to change direction quickly. The escapes should not be over-emphasized, however. The common names of several taxa of birds show their specialization in fly prey: gnatcatchers and flycatchers, for example.

Sit-and-wait predators such as spiders and mantises take a toll. Frogs and lizards, too, can make captures with their fast-moving tongues. Flies have fewer defenses against these predators, although the sheer numbers in many fly species mean that such predation as occurs is not dangerous to the survival of the species.

Flies are less likely than other insects to use camouflage and warning coloration, probably because their ability to fly away from predators means these other defenses are not so necessary. Chemical defenses also seem less common in Diptera than in most other orders.

Relationship to humans. A number of fly groups are vectors of human diseases. Among these maladies in Latin America are yellow fever, dengue fever, malaria, arbovirus encephalitis, leishmaniasis, and the zika virus disease. The role of flies in the spread of disease is so great, and so complex, that Bernard Greenberg wrote a two-volume, 1,303 page work on the subject in 1971.

Even when not carrying disease, certain kinds of flies can drive people to distraction. In 1925, Col. Percy Fawcett filed his last report from Brazil before his expedition lost contact with its sponsors and was never heard from again. Fawcett wrote, "The attempt to write is fraught with much difficulty owing to the legions of flies that pester one from dawn till dark—and sometimes all through the night!" These "stinging horrors get all over one's hands, and madden," Fawcett concluded.

Larval flies can cause catastrophic damage to crops by their feeding. While some adult and larval flies consume insect pests, they eat beneficial species as well. Farm animals are sometimes tormented by flies, and bites and the resulting wounds can be serious though usually not fatal.

Yet as the opening sentences of this section noted, flies provide many services to humans. Only hymenopterans pollinate more flowers than flies. Of equal importance is flies' processing of dung, carion, and rotting plant material into a safer and less odiferous material that becomes part of the topsoil used in agriculture.

Neotropical Specialties. Those who are already familiar with the European or North American fly fauna will notice some differences in the fauna of tropical America. Predictably, flies that flourish in cool environments, such as heleomyzid flies (Heleomyzidae) and root-maggot flies (Anthomyiidae) are scarcer. Other families are found primarily or exclusively in the American tropics, including Richardiidae, Ropalomeridae, and the giant timber flies that make up family Pantophthalmidae.

Figure 42 This *Willistoniella sp.* fly (Ropalomeridae) is feeding on papaya, with two nearby drosophilid fruit flies doing the same. *Willistoniella spp.* are found only in the Neotropics.

Tipulidae: Tipulinae
Nephrotoma sp. • Tiger Crane Fly
El Hato, Sacatepéquez, Guatemala

Limoniidae: Limoniinae
Dicranomyia sp. • Limoniid Crane Fly ♂
Nanegalito, Tandayapa valley, Ecuador

Limoniidae: Limoniinae
Hexatoma sp. • Limoniid Crane Fly
Parque Nacional Tayrona, Magdalena, Colombia

Limoniidae: Limnophilinae
≈ Epiphragma sp. • Limoniid Crane Fly
Along the Rewa River, Guyana

Limoniidae: Limoniinae
Trentepohlia sp. • Limoniid Crane Fly
Montsinéry, Montsinéry-Tonnegrande, French Guiana

Bibionidae: Bibioninae
Bibio sp. • March Fly
Boquete, Chiriquí, Panama

Limoniidae: Chioneinae
≈ Teucholabis sp. • Limoniid Crane Fly
Along the Rewa River, Guyana

Bibionidae: Bibioninae
Dilophus sp. • March Fly
Tuis, Cartago, Costa Rica

Keroplatidae: Keroplatinae
Orfeliini • Predatory Fungus Gnats
Along the Rupununi River, Guyana

Culicidae: Culicinae
Psorophora sp. • Mosquito
Filadelfia, Boquerón, Paraguay

Psychodidae: Psychodinae
Clogmia albipunctata • Drain Fly
Asunción, distrito Capital, Paraguay

Culicidae: Culicinae
Culicinae • Mosquito
Presqu'île de la Caravelle, Martinique

Culicidae: Culicinae
Psorophora cf. ferox • Mosquito
Along the Rewa River, Guyana

Simuliidae: Simuliinae
Simulium sp. • Black Fly
Naranjo, Alajuela, Costa Rica

Ceratopogonidae: Forcipomyiinae
Forcipomyia sp. • Biting Midge
Tuis, Cartago, Costa Rica

Ceratopogonidae: Forcipomyiinae
Forcipomyia sp. • Biting Midge
Tuis, Cartago, Costa Rica

Ceratopogonidae: Forcipomyiinae
Forcipomyia sp. • Biting Midges on a moth
Tuis, Cartago, Costa Rica

Chironomidae: Chironominae
≈ Chironomini • Non-biting Midge
Nueva Palmira, Colonia, Uruguay

Chironomidae: Chironominae
Chironomini • Green Midge
Suchitoto, Cuscatlán, El Salvador

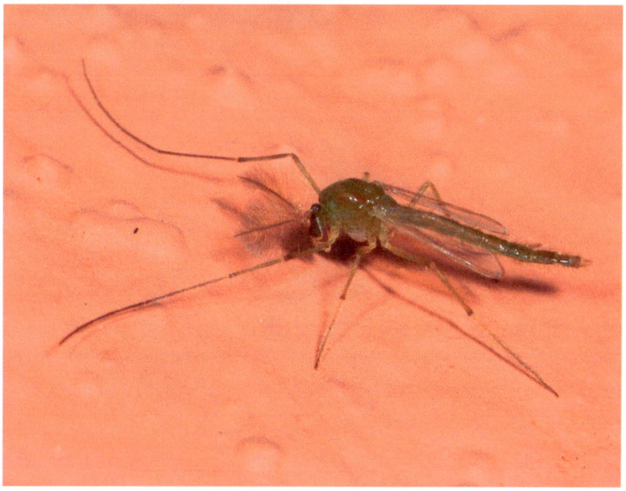

Chironomidae: Chironominae
Chironomini • Green Midge
Long Bay, Portland parish, Jamaica

Chironomidae: Tanypodinae
Tanypus (Apelopia) cf. neopunctipes • Non-biting Midge
Crooked Tree Preserve, Belize district, Belize

Stratiomyidae: Stratiomyinae
Nothomyia parvicornis • Small-horned Soldier Fly
Parque Nacional Pico Bonito, Honduras

Tabanidae: Tabaninae
Chlorotabanus inanis • Green Horse Fly
Montagne de Kaw, Roura, French Guiana

Stratiomyidae: Stratiomyinae
≈ Euparyphus sp. • Soldier Fly
Montagne de Kaw, Roura, French Guiana

Stratiomyidae: Beridinae
Heteracanthia cf. ruficornis • Soldier Fly
Suchitoto, Cuscatlán, El Salvador

Stratiomyidae: Stratiomyinae
≈ Hedriodiscus sp. • Soldier Fly
Montsinéry, Montsinéry-Tonnegrande, French Guiana

Stratiomyidae: Sarginae
Merosargus sp. • Soldier Fly
Suchitoto, Cuscatlán, El Salvador

Stratiomyidae: Pachygastrinae
Pachygaster group of genera • Soldier Fly
Tuis, Cartago, Costa Rica

Stratiomyidae: Sarginae
Microchrysa sp. • Soldier Fly
Along the Rupununi River, Guyana

Stratiomyidae: Clitellariinae
Diaphorostylus sp. • Soldier Fly
Tuis, Costa Rica

Stratiomyidae: Sarginae
Ptecticus sp. • Soldier Fly
Montsinéry, Montsinéry-Tonnegrande, French Guiana

Stratiomyidae: Clitellariinae
Euryneura sp. • Soldier Fly
Parque Nacional Tayrona, Magdalena, Colombia

Stratiomyidae: Clitellariinae
Cyphomyia cf. wiedemanni • Wiedemann's Soldier Fly
Novo Airão, Amazonas, Brazil

Stratiomyidae: Hermetiinae
Hermetia illucens • Soldier Fly
Novo Airão, Amazonas, Brazil

Stratiomyidae: Clitellariinae
Chordonota sp. • Soldier Fly
Filadelfia, Boquerón, Paraguay

Bombyliidae: Anthracinae
Ligyra sp. • Bee Fly
Selva Negra, Matagalpa, Nicaragua

Stratiomyidae: Hermetiinae
Hermetia sp. • Soldier Fly
Along the Rupununi River, Guyana

Bombyliidae: Lomatiinae
Macrocondyla setosa • Bee Fly
Puerto Madryn, Chubut, Argentina

Bombyliidae: Anthracinae
Villa sp. • Bee Fly
Jaragua National Park, Pedernales, Dominican Republic

Asilidae: Asilinae
Lochmorhynchus sp. • Robber Fly
Puerto Madryn, Chubut, Argentina

Asilidae: Asilinae
Apotinocerus cf. brevistylatus • Robber Fly
Puerto Madryn, Chubut, Argentina

Asilidae: Asilinae
Mallophora sp. • Bee Killer
Montsinéry, Montsinéry-Tonnegrande, French Guiana

Asilidae: Asilinae
Efferia nigrimystacea • Robber Fly
Presqu'île de la Caravelle, Martinique

Asilidae: Asilinae
Mallophora sp. • Bee Killer
Along the Rupununi River, Guyana

Asilidae: Asilinae
Prolatiforceps sp. • Robber Fly
El Hato, Sacatepéquez, Guatemala

Asilidae: Dasypogoninae
Diogmites sp. • Hanging-thief
Médanos de Punta Gorda, Colonia, Uruguay

Asilidae: Asilinae
Eicherax sp. • Robber Fly
Parque Nacional Tortuguero, Limón, Costa Rica

Asilidae: Stenopogoninae
Amblyonychus cf. trapezoidalis • Robber Fly
Parque Nacional Pico Bonito, Honduras

Asilidae: Dasypogoninae
Diogmites sp. • Hanging-thief
Tuis, Cartago, Costa Rica

Asilidae: Stenopogoninae
Amblyonychus sp. • Robber Fly
Berg en Dal, Brokopondo, Suriname

Asilidae: Stichopogoninae
Lissoteles sp. • Robber Fly
Paracas, Ica, Peru

Asilidae: Laphriinae
Atomosia cf. tibialis • Laphriine Robber Fly
Crooked Tree Preserve, Belize district, Belize

Asilidae: Laphriinae
Atomosia sp. • Laphriine Robber Fly
Filadelfia, Boquerón, Paraguay

Asilidae: Laphriinae
Atoniomyia sp. • Laphriine Robber Fly
Tuis, Cartago, Costa Rica

Asilidae: Laphriinae
Eumecosoma sp. • Laphriine Robber Fly
Tuis, Cartago, Costa Rica

Asilidae: Laphriinae
Pilicata sp. • Laphriine Robber Fly
Valle de Anton, Coclé, Panama

Therevidae: Therevinae
Penniverpa cf. epidema • Stiletto Fly
Crooked Tree Preserve, Belize district, Belize

Dolichopodidae: Dolichopodinae
Paraclius sp. • Long-legged Fly
Novo Airão, Amazonas, Brazil

Empididae: Empidinae
Hilarempis sp. • Dance Fly
Reserva Forestal Magallanes, Punta Arenas, Chile

Dolichopodidae: Dolichopodinae
Paraclius sp. • Long-legged Fly
Colakreek, Para, Suriname

Hybotidae: Hybotinae
Hybos sp. • Hybotid Dance Fly
El Hato, Sacatepéquez, Guatemala

Dolichopodidae: Medeterinae
Prob. Medetera sp. • Long-legged Fly
Selva Negra, Matagalpa, Nicaragua

Dolichopodidae: Plagioneuriinae
Plagioneurus univittatus • Long-legged Fly
Crooked Tree Preserve, Belize distirct, Belize

Dolichopodidae: Sciapodinae
Condylostylus sp. • Long-legged Fly
Novo Airao, Amazonas, Brazil

Dolichopodidae: Sciapodinae
Condylostylus sp. • Long-legged Fly
El Hato, Sacatepéquez, Guatemala

Dolichopodidae: Sciapodinae
Condylostylus sp. • Long-legged Fly
Montsinéry, Montsinéry-Tonnegrande, French Guiana

Dolichopodidae: Sciapodinae
Condylostylus sp. • Long-legged Fly
Santo Domingo de los Colorados, Pichincha, Ecuador

Dolichopodidae: Sciapodinae
Amblypsilopus sp. • Long-legged Fly
Boquete, Chiriquí, Panama

Dolichopodidae: Diaphorinae
Diaphorus sp. • Long-legged Fly
Selva Negra, Matagalpa, Nicaragua

Dolichopodidae: Diaphorinae
Diaphorus sp. • Long-legged Fly
Berg en Dal, Brokopondo, Suriname

Phoridae: Metopininae
Megaselia sp. • Humpbacked Fly
Boquete, Chiriquí, Panama

Syrphidae: Eristalinae
Copestylum sp. • Snout Fly
Colonia del Sacramento, Colonia, Uruguay

Syrphidae: Eristalinae
Eristalis tenax • Drone Fly
Puerto Madryn, Chubut, Argentina

Syrphidae: Eristalinae
Ornidia cf. obesa • Obese Syrphid Fly
Novo Airão, Amazonas, Brazil

Syrphidae: Eristalinae
Palpada vinetorum • Syrphid Fly
Manaus, Amazonas, Brazil

Syrphidae: Syrphinae
Hybobathus aff. rubricosus • Syrphid Fly
Shintuya Hot Springs, Madre de Dios, Peru

Syrphidae: Eristalinae
Palpada sp. • Syrphid Fly
Montagne de Kaw, Roura, French Guiana

Syrphidae: Syrphinae
Hybobathus norina • Syrphid Fly
Shintuya Hot Springs, Madre de Dios Department, Peru

Syrphidae: Eristalinae
Sterphus (Crepidomyia) plagiatus • Syrphid Fly
Shintuya Hot Springs, Madre de Dios Department, Peru

Syrphidae: Syrphinae
Ocyptamus antiphates • Syrphid Fly
Berg en Dal, Brokopondo, Suriname

Syrphidae: Syrphinae
Ocyptamus cf. luctuosus • Syrphid Fly
Berg en Dal, Brokopondo, Suriname

Syrphidae: Syrphinae
Toxomerus cf. dispar • Syrphid Fly
El Hato, Sacatepéquez, Guatemala

Syrphidae: Syrphinae
Toxomerus sp. • Syrphid Fly
El Hato, Sacatepéquez, Guatemala

Micropezidae: Taeniapterinae
Cardiacephala brevipennis • Stilt-legged Fly
Berg en Dal, Brokopondo, Suriname

Syrphidae: Syrphinae
Toxomerus nr. politus • Syrphid Fly
Acjanaco, Manú National Park, Peru

Micropezidae: Taeniapterinae
Cardiacephala sp. • Stilt-legged Fly
Parque Nacional Tayrona, Magdalena, Colombia

Micropezidae: Taeniapterinae
Grallipeza cliffi • Stilt-legged Fly
Long Bay, Portland parish, Jamaica

Micropezidae: Taeniapterinae
Plocoscelus sp. • Stilt-legged Fly
Réserve Naturelle Trésor, Roura, French Guiana

Micropezidae: Taeniapterinae
Grallipeza nr. ecuadorensis • Stilt-legged Fly
Reserva Particular do Cristalino, Mato Grosso, Brazil

Micropezidae: Taeniapterinae
Plocoscelus sp. • Stilt-legged Fly
Parque Nacional Tayrona, Magdalena, Colombia

Micropezidae: Taeniapterinae
≈ Paragrallomyia sp. • Stilt-legged Fly
Novo Airão, Amazonas, Brazil

Micropezidae: Taeniapterinae
Ptilosphen variolatus • Stilt-legged Fly
Selva Negra, Matagalpa, Nicaragua

Micropezidae: Taeniapterinae
Poecilotylus sp. • Stilt-legged Fly
Réserve Naturelle Trésor, Roura, French Guiana

Neriidae
Glyphidops cf. filosus • Cactus Fly
Berg en Dal, Brokopondo, Suriname

Micropezidae: Taeniapterinae
Taeniaptera sp. *sensu lato* • Stilt-legged Fly
Suchitoto, Cuscatlán, El Salvador

Neriidae
Glyphidops cf. flavifrons • Cactus Fly
Crooked Tree Preserve, Belize district, Belize

Micropezidae: Micropezinae
Micropeza sp. •Stilt-legged Fly
Along Arroyo de las Víboras, Colonia, Uruguay

Richardiidae: Richardiinae
Richardia sp. • Richardiid Fly
Novo Airão, Amazonas, Brazil

Ulidiidae: Ulidiinae
Euxesta sp. • Corn Silk Fly
Selva Negra, Matagalpa, Nicaragua

Tephritidae: Tephritinae
Rhynencina sp. • Fruit Fly
Paracas, Ica Region, Peru

Tephritidae: Trypetinae
Anastrepha sp. • Fruit Fly
Rewa village, Upper Takutu-Upper Essequibo, Guyana

Pyrgotidae
≈ Sphecomyiella sp. • Light Fly
Selva Negra, Matagalpa, Nicaragua

Tephritidae: Tephritinae
Tetreuaresta bartica • Fruit Fly
Montagne de Kaw, Roura, French Guiana

Lauxaniidae: Lauxaniinae
≈ Melanomyza sp. • Lauxaniid Fly
Montsinéry, Montsinéry-Tonnegrande, French Guiana

Lauxaniidae: Lauxaniinae
Allominettia maculatifrons • Lauxaniid Fly
Along the Rewa River, Guyana

Lauxaniidae: Lauxaniinae
Hirtodeceia sp. • Lauxaniid Fly
Berg en Dal, Brokopondo, Suriname

Lauxaniidae: Lauxaniinae
Baliopteridion sp. • Lauxaniid Fly
Along the Rupununi River, Guyana

Lauxaniidae: Lauxaniinae
≈ Minettia sp. • Lauxaniid Fly
Berg en Dal, Brokopondo, Suriname

Lauxaniidae: Lauxaniinae
Chaetominettia corollae • Lauxaniid Fly
Berg en Dal, Brokopondo, Suriname

Lauxaniidae: Lauxaniinae
Neogriphoneura timida • Lauxaniid Fly
Parque Nacional Tayrona, Magdalena, Colombia

Lauxaniidae: Lauxaniinae
Poecilominettia effosa species group • Lauxaniid Fly
Boquete, Chiriquí, Panama

Sciomyzidae: Sciomyzinae
Protodictya sp. • Marsh Fly
Filadelfia, Boquerón, Paraguay

Lauxaniidae: Lauxaniinae
Poecilominettia sexseriata • Lauxaniid Fly
Rewa village, Upper Takutu-Upper Essequibo, Guyana

Sciomyzidae: Sciomyzinae
Thecomyia sp. • Marsh Fly
Along the Rewa River, Guyana

Lauxaniidae: Lauxaniinae
Siphonophysa sp. • Lauxaniid Fly
Along the Rupununi River, Guyana

Ropalomeridae
Willistoniella sp. • Ropalomerid Fly
Parque Nacional Pico Bonito, Atlántida, Honduras

Sepsidae: Sepsinae
≈ Microsepsis sp. • Black Scavenger Fly
Berg en Dal, Brokopondo, Suriname

Heleomyzidae
Prosopantrum sp. • Heleomyzid Fly
Reserva Forestal Magallanes, Punta Arenas, Chile

Periscelididae: Periscelidinae
Genus Neoscutops, Marbenia, or Scutops
Montsinéry, Montsinéry-Tonnegrande, French Guiana

Curtonotidae: Curtonotinae
Curtonotum sp. • Hunchbacked Fly
Colakreek, Para, Suriname

Periscelididae: Stenomicrinae
Stenomicra sp. • Sap Fly
Selva Negra, Matagalpa, Nicaragua

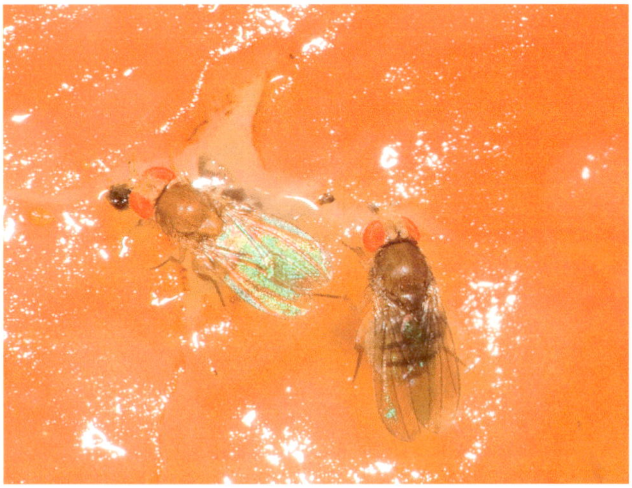

Drosophilidae: Drosophilinae
Drosophila sp. • Pomace Fly
Along the Rio Negro, Amazonas, Brazil

Drosophilidae: Drosophilinae
≈ Hirtodrosophila sp. • Pomace Fly
Leona Vicario, Quintana Roo, Mexico

Anthomyiidae: Anthomyiinae
Anthomyia sp. • Root-Maggot Fly
Reserva Forestal Magallanes, Punta Arenas, Chile

Drosophilidae: Drosophilinae
Zaprionus indianus • African Pomace Fly
Manaus, Amazonas, Brazil

Anthomyiidae: Anthomyiinae
Fucellia maritima • Seaweed Fly
Shore along Magellan Strait, Punta Arenas, Chile

Scathophagidae: Scathophaginae
Scathophaga tropicalis • Tropical Dung Fly
Copacabana, La Paz department, Bolivia

Muscidae: Muscinae
Musca domestica • House Fly
Paracas, Ica, Peru

Muscidae: Azeliinae
Palpibracus sp. • Muscid Fly
Reserva Forestal Magallanes, Punta Arenas, Chile

Sarcophagidae: Sarcophaginae
Poss. Dexosarcophaga sp. • Flesh Fly
Montagne de Kaw, Roura, French Guiana

Calliphoridae: Toxotarsinae
Sarconesia sp. • Blow Fly
Yumani, beside Lago Titicaca, Bolivia

Tachinidae: Tachininae
≈ Trichophora sp. • Tachinid Fly
Selva Negra, Matagalpa, Nicaragua

Sarcophagidae: Sarcophaginae
Microcerella sp. • Flesh Fly
Isla del Sol, Lago Titicaca, Bolivia

Tachinidae: Exoristinae
Blondeliini • Tachinid Fly
Sentier de Bagne des Annamites, French Guiana

Tachinidae: Phasiinae
≈ Trichopoda sp. • Tachinid Fly
Berg en Dal, Brokopondo, Suriname

Tachinidae: Dexiinae
Cordyligaster cf. petiolatus • Tachinid Fly
Parque Nacional Pico Bonito, Atlántida, Honduras

Tachinidae: Dexiinae
Cholomyia cf. inequipes • Tachinid Fly
Montsinéry, Montsinéry-Tonnegrande, French Guiana

Asilidae: Stichopogoninae
Prolepsis lucifer • Lucifer Robber Fly
Zagarzazú, Colonia, Uruguay

10

ORDER TRICHOPTERA

Caddisflies

Caddisflies are the architects and builders *par excellence* of the insect world. Other insect orders have their own achievements, of course. Groups of Hymenoptera build impressive structures together—termites too—but their building materials seem dubious from the human perspective: feces, saliva, mud, and masticated wood. In the case of Hymenoptera, the work of building falls to the mature insects, while the young larvae are helpless and require the attention of the busy workers. Caddisflies, in many families, are building their first structure within an hour of hatching, using rock and sand and wood and some incredibly strong silk. Hatchlings of other families of caddisfly quickly set to work to build huge nets (relative to their body size), also of strong silk, and they expertly position it to begin catching prey immediately.

Each species of caddisfly has its own type of shelter ("case") or net, and these can be useful in making identifications. But note that a few groups of caddisfly larvae are free-living and must hide among plants or in crevices to escape predation.

Caddisflies are related to moths, and adult caddisflies are often confused with Lepidoptera. There are several key differences between the two groups, however. Lepidoptera wings are covered with scales, but in Trichoptera the wings are covered with hairs (sometimes there are some scales too, especially along the veins). Lepidoptera have a coiled tube for a mouth, but no Trichoptera have that feature, and adult caddisflies feed lightly if they feed at all. Finally, caddisflies have long antennae, often much longer than their body or wings, while Lepidopteran antennae are much shorter.

Like the wings, the bodies of caddisflies are hairy. The compound eyes are large, and two or three ocelli are present. Caddisflies hold their wings over their bodies like a peaked roof, though in some species the wings are only slightly "peaked."

Nearly every species of caddisfly lives in or beside a rivulet, stream, river, pond, or lake. The few that do not, live in areas of very wet soil such as a seep or swamp. Most adult caddisflies are crepuscular or nocturnal.

Trichoptera contains about 45 families worldwide, and approximately 14,000 named species. The Neotropical list contains about 2,000 species. The name Trichoptera comes from the Greek and means "hairy wings." The common name caddisfly is related to "caddis men" of the Middle Ages, fabric merchants who pinned squares of cloth to their clothing as samples of their wares, just as caddisfly larvae often have a case that looks like patchwork.

Food. Caddisfly larva feed mostly on plants, including fragments of algae, mosses, and other water plants. Some species are predaceous and capture prey with their forelegs. Species that use nets seem to feed primarily on tiny animals they catch, but their nets may provide them with plant-based food as well. Sometimes the net-building larvae behave like spiders, as they rush to trapped prey and subdue it with silk.

Adult caddisflies have reduced mouthparts, and many do not feed. Some adults are seen feeding on flowers, presumably on the pollen found there.

Reproduction. Males and females locate each other primarily by sight in some families. In others, males home in on female-produced pheromones. Males sometimes swarm together, possibly to make a bigger target for females to find. Impressively, mating is usually accomplished in mid-air.

Females deposit their eggs in water, sometimes actually entering the water to oviposit. In some cases, eggs are laid above the water on an overhanging rock so that when the nymphs hatch they will drop into the water. Eggs are typically laid in clusters, and many are lost to fish and hungry aquatic invertebrates.

Development. Most caddisfly larvae go through five instars, while a few species go through six or seven. The larvae have abdominal gills and can also take in oxygen through the cuticle. They lack the several pairs of abdominal prolegs possessed by caterpillars, but typically have a pair of prolegs at the apex of the abdomen. Some larvae use the prolegs to hold themselves in their case. Silk is also used for that purpose.

The ending of the pupal stage is unusual. The pupa splits its silky cocoon and the pupa that emerges has freely moving legs modified for swimming. It swims to dry land, then splits the pupal "skin" and emerges as an adult.

Dangers to caddisflies. Fish of various sizes are the primary predators of larval caddisflies, though frogs and salamanders and various invertebrates also take their toll. Among the latter are the larvae of dragonflies, stoneflies, and dobsonflies. Some larval caddisflies will eat smaller larval caddisflies of the same or a different species. Adults confront the usual dangers faced by terrestrial and aerial insects, namely predation by birds, reptiles, amphibians, bats, spiders, and predacious insects.

Parasites including mites and protozoans attack caddisfly larvae, and the pupae are attacked by parasitoids including several genera of ichneumonid wasps.

Defenses. The primary defenses of larvae are, of course, the structures they have constructed. The cases are well-built, and the larva can retreat entirely inside. The nets constructed by some caddisfly groups also serve as a retreat from predators.

Adults experience reduced bird predation by being active primarily at night.

Relationship to humans. Trichoptera do not have an important economic relationship to humans.

Anglers have used caddisfly larva for bait for centuries, particularly in trout fishing. The larvae, pupae, and adults have also been modeled in tying flies. Many caddisflies have common names devised by anglers. For example, caddisflies in the genus *Pycnopsyche* are called Great Autumn Brown Sedges.

Trichoptera are also used as bioindicators of a stream's health. Many species require extremely clean water, for example, and the disappearance of such a species from a steam is an indication of declining water quality.

Neotropical Specialties. Two particularly nice species of adult caddisflies in the Neotropics boast attractive colors and patterns. One is *Macronema hageni*, found most often in in Brazil and Peru. The other is *Macrostemum arcuatum*, with records from Brazil and Guyana. Photographers can get some nice photos of these pretty species, but unfortunately for collectors the attractive colors fade to drab in pinned specimens.

Figure 43 Though it resembles the engravings of an earlier era, this depiction of scenes from the life history of caddisflies is from 1957 and is by the artist Claus Caspari.

Hydropsychidae: Macronematinae
Leptonema amazonense group • Net-spinning Caddisflies
Colakreek, Para, Suriname

Hydropsychidae: Macronematinae
Leptonema cf. albovirens • Net-spinning Caddisflies
Tuis, Cartago, Costa Rica

Hydropsychidae: Macronematinae
Leptonema sp. • Net-spinning Caddisflies
Selva Negra, Matagalpa, Nicaragua

Hydropsychidae: Macronematinae
Macronema hageni • Net-spinning Caddisflies
Novo Airão, Amazonas, Brazil

Hydropsychidae: Macronematinae
Macrostemum arcuatum • Net-spinning Caddisfly
Rewa village, Upper Takutu-Upper Essequibo, Guyana

Hydropsychidae: Macronematinae
Synoestropsis sp. • Net-spinning Caddisflies
Rewa village, Upper Takutu-Upper Essequibo, Guyana

Leptoceridae: Leptocerinae

Oecetis sp. • Longhorned Caddisfly

Crooked Tree Preserve, Belize district, Belize

Leptoceridae: Leptocerinae

Nectopsyche acutiloba • Longhorned Caddisfly

Colakreek, Para, Suriname

Odontoceridae: Odontocerinae

Marilia cf. flexuosa • Mortarjoint Casemaker

Rewa village, Upper Takutu-Upper Essequibo, Guyana

11

ORDER LEPIDOPTERA

Moths and Butterflies

Butterflies and moths are more alike than different. Spanish-speakers use the same word for both—mariposas. Moths are referred to as mariposas "de la noche" (of the night). Indeed, one chief difference between these two groups is behavioral: butterflies are active in the daytime, and moths at night—though there are exceptions.

Figure 44 Maria Sibylla Merian's portrait of a blue morpho *(Morpho helenor)* showing the upper sides of the wings. Circa 1700, Suriname.

Most butterflies have clubbed antennae while moths have threadlike or feathery antennae. Members of the butterfly family Hesperiidae (skippers) have hooked antennae. Another structural difference is that in moths the forewings are connected to the hindwings by hooks, while in butterflies the forewings and hindwings are not attached to each other. This latter difference is not usually helpful when studying Lepidoptera with binoculars or a camera, however.

Moths' and butterflies' similarities are many. Their mouth-parts are tubes used for sucking liquid food. These are coiled underneath the head when not in use. Covering the body and wings are flattened scales called setae. It is these scales that provide the moths' and butterflies' color and markings.

Taxonomy. The order Lepidoptera comprises 125 families worldwide. Of these, six families make up a monophyletic group called superfamily Papilionoidea. (Monophyletic means all members in the group are descended from a single common ancestor.) The English names for these butterfly families are skippers, swallowtails, metalmarks, hairstreaks, brushfooted butterflies, and the whites and yellows. The seventh and last family of butterflies, Hedylidae, is placed in its own superfamily, Hedyloidea. This family has the common name American moth-butterflies. As the common name suggests, these butterflies have some traits more common to moths, such as thread-like antennae. Family Hedylidae is found only in the Neotropics, and contains only one genus, *Macrosoma*.

Thus seven families of Lepidoptera are butterflies and the other 118 families bear the common name moth. There is no monophyletic scientific name for the moths group.

Food. Lepidoptera larvae (caterpillars) are voracious. Many feed on leaves. Other larval food sources, in some groups, include seeds, flowers, stored grains, wood, and even cloth.

The food sources for adult members of this order are varied. Nectar is the primary food for a great many species. Some also consume pollen or the juices of rotting fruit. Some adult Lepidoptera have no mouthparts, or only vestigial ones, and do not feed. Among these are some of the giant silk moths (Saturniidae), which are widespread in tropical America. Eight species of moth in Erebidae: Calpinae have been documented feeding on the blood of humans or other mammals, but none of these so-called vampire moths are found in the Neotropics.

Butterflies in particular may need to supplement their diet with minerals lacking in their primary food source. They therefore engage in puddling behavior, gathering to feed either at a muddy puddle or on other mineral sources such as dung, carrion, or a sweaty human. Soil soaked with urine is also attactive to butterflies. In some species it has been shown that males are more likely to puddle and that their puddling increases their reproductive success. Moths sometimes puddle too, but are observed less often because they are active at night.

Reproduction. Males and females of many species of moth locate each other using chemical signals called pheromones. Females emit the pheromone and males move towards the source. Males of some groups can detect the pheromone at distances of up to several miles. To aid them in their quest, males sometimes have feathery antennae, which detect the pheromone, while the females of the same species have simpler, thread-like antennae. As for butterflies, the females of a few species have glands that produce attractant pheromones, but the great majority do not.

Figure 45 Butterflies of several species (primarily in Pieridae) are engaged in puddling on the banks of the Cristalino River, Brazil.

Figure 46 At the end of the abdomen of this male crambid moth *(Diaphania hyalinata)* are "hair pencils." After the female moth releases her sexual pheromones, a response is triggered in the male, and the internal hair pencils are everted from his abdominal apex. The pencils help disperse the male sexual pheromone.

Some moths and most butterflies use a method other than pheromones to enable the two sexes to find each other. One reason for the bright color and markings in most butterflies and some moths is that they help males and females of the same species find each other. Wing shape and patterns of movement can also help the sexes locate each other and ensure that both are of the same species. Note that in many species the males and females look nearly identical, while in others the two sexes have different colors and markings.

A few species of butterflies and moths use sounds as a part of their courting. Unfortunately for the naturalist, humans can't hear these wooing songs of male lepidopterans, both because the volume is low (presumably so bats won't hear and attack) and because the frequency is above that detectable by humans. Among the species that employ sound in courtship is *Achroia grisella,* the Lesser Wax Moth, found in many warmer areas of the Neotropics.

Figure 17 Often male moths have feathery or comb-like antennae to allow them to detect the females' sexual attractant pheromone. Woolly moths like these have acquired the name "poodle moth" on the internet, and this one is likely *Titya cinerascens* (Lasiocampidae). French Guiana.

Lepidoptera mate in a tail to tail position. After mating, the females lay eggs on the larval food sources, frequently the leaf of a plant, or in some cases in stored grains, on carpets, or on other larval foodstuffs. Note that in many Lepidoptera species the caterpillars can develop only on a single species of plant, or a small group of related plants. Thus it is crucial that the egg-laden female select the right location for ovipositing—the caterpillars' survival would be reduced if they had to spend days searching for food at their relatively slow gait. For the naturalist, identifying the food plant can be helpful in identifying the insect.

Caterpillars. The eggs typically hatch after a period of weeks or months, depending on the species and on environmental conditions. In some species the hatchlings eat their own eggshell, obtaining nutrients in this way. Other species find it more advantageous for the caterpillar to begin eating the primary larval food source immediately.

Most caterpillars have a hardened head capsule, and a softer thorax with a pair of legs attached to each of three thoracic segments. The legs tend to have a conical shape and to curve inwards toward the body. The caterpillar's soft abdomen typically bears ten discernible segments, and some of these segments bear prolegs, which are fleshy protuberances that aid in locomotion. While there is considerable variety from family to family or even from species to species within a family, a majority of caterpillars have prolegs on abdominal segments three to six, and on segment ten.

Figure 48 Like all Lepidoptera, these metalmarks *(Detritivora barnesi)* mate in a tail-to-tail position. Emberá-Wounaan indigenous area, Panama.

The primary function of caterpillars is to eat. As the caterpillar grows, it must molt periodically. Most caterpillars go through four or five instars, but some have more. The period of molting is a dangerous one for caterpillars, as they need to be relatively motionless for hours, and may be unable to deploy any defenses while their body is half-encased in the old "skin" during the molting process.

One thing that makes identifying caterpillars difficult is that their appearance can change markedly from instar to instar. The changes can involve color, markings, or the appearance and disappearance of horns, hairs, spines, and the like. A thorough field guide to caterpillars would need five or more photos for each species.

Pupae. After the caterpillar has passed through the requisite number of instars for its species, it undergoes a final larval molt into a pupa. The pupal stage is a period of transformation into the adult. Pupae are very often attached to twigs, but some species attach themselves to some other surface, or pass the pupal stage underground. Some pupae are hidden under stones, beneath bark, or in crevices.

The butterfly pupa is normally naked or fleshy in appearance, and is often called a chrysalis. This word derives from the Greek chrysós, meaning gold—some chrysalises such as that of the monarch have gold markings. Many moths construct a silken cocoon to shelter themselves in the pupal stage. The silk is sometimes mixed with larval urticating setae (stinging hairs) to give it an added protective quality, or mixed with plant parts that will help camouflage it.

Most Lepidoptera are nearly motionless during the pupal stage, but some can make a jerking motion that helps them get away from a predator. Pupae that live in the ground, or under bark, are able to wriggle their way to a place where there is room enough to molt the pupal exoskeleton.

After a pupal period that varies considerably in length among the various groups, the adult moth or butterfly emerges. The emergence, called eclosion, is another period of danger for the insect, as its wings are not functional until they fill with haemolymph, and the body itself is soft. It can take a few hours for the wings to fill and the cuticle to harden.

Broods. The life history of a certain species of moth or butterfly can vary considerably because of climatic conditions. In a cool location such as southern Chile or Argentina, the species may produce one or two broods per year. In a warmer location, the same species might produce three or four broods. In some species capable of rapid development and living in a warm and humid location (such as the Amazon or the Guianas) eight broods per year are possible.

The time needed for development from egg to adult varies considerably among species. Cold weather species, such as those found in the Andes, may take two or more years for a single brood. The same is true of species that live in very dry locations such as coastal Peru. In both cases, the species undergo a period of diapause, essentially a resting period that lasts until better temperatures or moisture levels return. Most commonly, the diapause occurs during the pupal period, but there are a number of exceptions. Long development times are also needed for species with a diet low in nutrients. For example, the larvae of Carpenter Moths (Cossidae) feed on wood, which is difficult to digest for most kinds of animal life. A development time of three years is typical for members of this family.

Perils for Leipdoptera. In all four stages of their lives, moths and butterflies face a number of dangers. The most obvious is predation. Birds, bats, spiders, lizards, frogs, and insectivorous insects are the creatures that most commonly feed on Lepidoptera.

All stages also face the danger of parasitoids—caterpillars most of all. In a typical case of parasitoidism, a tachinid fly or an ichneumonid wasp lays eggs on the outer surface of the caterpillar, or injects the eggs into the caterpillar's body. When those eggs hatch, the larval parasitoid begans feeding on the caterpillar, often avoiding the vital organs in order to keep their host alive. Eventually the larval parasitoids mature and leave the dead or dying caterpillar's body through an exit hole.

While the actions of parasitoids result in the death of the host, those of parasites do not. One common type of parasite found on adult and larval Lepidoptera are biting midges in the genus *Forcipomyia*. These flies perch on a caterpillar or adult lepidopteran, then use their piercing mouthparts to feed on their host's haemolymph. The flies can swell to resemble a tiny balloon or a swollen tick. In fact, they have the common name stick tick. Mites (often bright red) are another common form of blood-sucking parasite, but in some cases the mites are harmless to their host and are just hitching a ride to a new location. Pseudoscorpions also use adult moths and butterflies as transportation. The term for this ride-taking is phoresy.

Disease is a life-threatening phenomenon for Lepidoptera. Causes of illness in butterflies and moths include viral, bacterial, and fungal infections. A common sight on a walk in a Neotropical forest is a moth covered with fungus. The fungus is usually light-colored, and may appear as spots, a sheet, or raised "fingers." Because they specialize in attacking insects, the fungi are termed entomopathogenic. Among the genera of fungi that attack Lepidoptera are *Cordyceps* and *Nosema*.

Defenses. Of course, moths and butterflies have defenses against various kinds of attack. Some of these defenses are quite passive, such as seeking a sheltered place or using camouflage to blend in with the substrate.

On a more active level, both caterpillars and adults may drop to the forest floor to get away from a predator. There they blend in, or hide under a dead leaf. Caterpillars may descend on a silk thread to make their fall safer, and may be able to reascend on the same thread when the danger has passed. Playing dead is another tactic employed by some taxa, such as moths in genus *Spilosoma* in family Erebidae.

Needless to say, flying away is a prime defense for adults, and they may employ a zigzag flight to make capture more difficult. Even if captured, slippery scales may allow an escape. Some adults have stout leg spurs they can use to wound an attacker.

Many Lepidoptera possess chemical defenses. These may be active defenses, where a caterpillar sprays a noxious compound or vomits or "reflex bleeds" one. The defense can also be passive, as when an adult or larva is poisonous or unpalatable to those trying to eat it. The poisons usually derive from the larval food plant. Either the species has evolved to be unaffected by the plant's poisons, or it safely sequesters the poison in a special chamber of the body.

To be most effective as a defense, the poisons must be advertised by bright, warning colors, often red or orange and black. Thus a bird, for example, will learn after one nauseating attempt not to attack that kind of caterpillar with the red spots and black stripes. The warnings are even more effective if a number of noxious species share the same basic pattern of warning colors and markings. This is called Mullerian mimicry.

Another type of mimicry is called Batesian, after the famous naturalist Henry Walter Bates, who developed the concept while observing and collecting butterflies and beetles in the Amazon region of Brazil. In Batesian mimicry species that are chemically harmless "adopt" (through evolution) the colors and patterns of a chemically protected species in order to discourage predators.

Some species of Lepidoptera mimic other kinds of insects, most often aggressive species of wasp. Others (larvae or adults) immitate something not usually sought as a foodstuff, such as bird droppings.

Large eyespots may make a lepidopteran look more formidible than it is. If the eyespots are hidden on the hindwing they may suddenly be flashed to startle the predator. Eyespots on the wings may encourage the attacker to bite the wing (thinking it is the head) instead of the more vulnerable body of the moth or butterfly. Oversized eyespots on larvae make the caterpillar seem like a larger and potentially dangerous creature.

Figure 49 This tiger moth, *Corematura chrysogastra*, undoubtedly frightens predators with its aposematic colors and its mimicry of a wasp.

As a defense against cold some Lepidoptera take refuge under bark or in some other sheltered place. Moths that are too cold to fly respond to the situation by shivering, which often warms their muscles sufficiently for flight. The monarch *(Danaus plexippus)* migrates from the chilly climes of eastern Canada and the U.S. to shelter in fir trees in central Mexico. On the return trip monarchs reproduce in several broods, thus the southward migrants never make it back home, but their decendants do.

Neotropical specialties. Several unusual genera are found in Central and South America. One is Urania, a beautiful large moth colored in green, black, and white that resembles a butterfly in some ways. It is active in daytime, and like some common butterflies it has swallow-tails. Unlike butterflies, it boasts thread-like antennae, and unlike butterflies it is attracted to lights. Urania is also worth remarking upon because at least one member of the genus is migratory.

Figure 50 Monarchs rest up for the northward journey at El Rosario Monarch Butterfly Preserve in Mexico's Michoacán state.

Another interesting genus, *Hamadryas* (common name: crackers) are noteable for producing a sound that resembles an electrical short. Males sally out from their tree trunk perch and emit this sound, either as a part of a territorial display aimed at other males or to make a favorable impression on females.

Two closely related Neotropical genera are noted for their large size and their beauty. These are the morphos *(Morpho spp.)* and owls *(Caligo spp.)*. The owls have more subtle coloration, in pleasing shades of brown, tan, and yellow, and sometimes smaller areas of red or blue. The morphos have similar subtle coloration on the underside, but are a bright blue color on the upper surfaces. When they fly they seem especially large and colored like nothing else in the forest.

Figure 51 This lepidopteran perched on a tree trunk in Brazil has fallen victim to an entomopathogenic fungus.

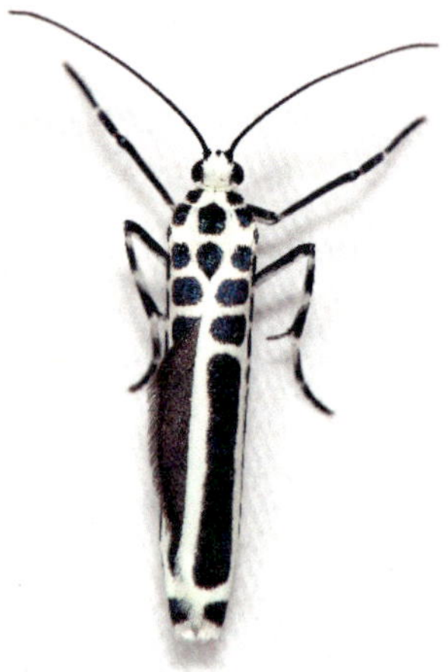

Attevidae: Attevinae
Atteva zebra • Zebra Atteva Moth
Tuis, Cartago, Costa Rica

Choreutidae: Choreutinae
Tortyra argentifascia • Metalmark Moth
Playa Los Angeles, Magdalena, Colombia

Pterophoridae
Pterophoridae • Plume Moth
El Hato, Sacatepéquez, Guatemala

Choreutidae: Choreutinae
Tortyra argentifascia • Silver-banded Metalmark Moth
Leona Vicario, Quintana Roo, Mexico

Choreutidae: Brenthiinae
Brenthia monolychna • Metalmark Moth
Playa Los Angeles, Magdalena, Colombia

Tortricidae: Tortricinae
Mictopsichia cf. benevides • Tortricid Leafroller moth
Along the Rupununi River, Guyana

Tortricidae: Olethreutinae
Cacocharis cymotoma • Tortricid Leafroller moth
Presqu'île de la Caravelle, Martinique

Tortricidae: Olethreutinae
Omiostola sp. • Tortricid Leafroller moth
Réserve Naturelle Trésor, Roura, French Guiana

Tortricidae: Olethreutinae
Dichrorampha sp. • Tortricid Leafroller moth
Salvador, Manú National Park, Peru

Tortricidae: Chlidanotinae
Auratonota petallocrossa • Tortricid Leafroller moth
Tuis, Cartago, Costa Rica

Tortricidae: Chlidanotinae
Hilarographa plectanodes • Torticid Leafroller moth
Tuis, Cartago, Costa Rica

Tortricidae: Chlidanotinae
Thaumatographa sp. • Tortricid Leafroller moth
El Hato, Sacatepéquez, Guatemala

Cossidae: Zeuzerinae
Morpheis sp. • Carpenter Moth
Mountain Pine Ridge Forest Reserve, Belize

Cossidae: Hypoptinae
Givira sp. • Carpenter Moth
Parque Nacional Pico Bonito, lodge, Atlántida, Honduras

Castniidae: Castniinae
Castniomera atymnius • Castniid Moth
Playa Los Angeles, Magdalena, Colombia

Castniidae: Castniinae
Telchin licus • Castniid Moth
Montsinéry, Montsinéry-Tonnegrande, French Guiana

Sesiidae: Sesiinae
Melittia sp. • Vine Borer
Montagne de Kaw, Roura, French Guiana

Sesiidae: Sesiinae
Sesiinae • Clearwing Moth
La Suiza, Cartago, Costa Rica

Dalceridae: Acraginae
Acraga sp. • Dalcerid Moth
Berg en Dal, Brokopondo, Suriname

Limacodidae: Limacodinae
Parasa viridogrisea • Green-gray Slug Moth
Sansón, Darién, Panama

Limacodidae: Limacodinae
Epiperola grandiosa • Grandiose Slug Moth
Berg en Dal, Brokopondo, Suriname

Limacodidae: Limacodinae
Perola villosipes • Slug Moth
Montagne de Kaw, Roura, French Guiana

Limacodidae: Limacodinae
Miresa clarissa • Slug Moth
Montagne de Kaw, Roura, French Guiana

Megalopygidae: Megalopyginae
Megalopyge lanata • Wooly Flannel Moth
Parque Nacional Pico Bonito, lodge, Atlántida, Honduras

Megalopygidae: Megalopyginae
Megalopyge albicollis • Flannel Moth
Tuis, Cartago, Costa Rica

Pyralidae: Chrysauginae
Semnia auritalis • Snout Moth
Tuis, Cartago, Costa Rica

Megalopygidae: Megalopyginae
Trosia nigropunctigera • Rosy Flannel Moth
Tuis, Cartago, Costa Rica

Pyralidae: Chrysauginae
Tamyra ignitalis • Fiery Bell Moth
Leona Vicario, Quintana Roo, Mexico

Zygaenidae: Procridinae
Neoprocis floridana • Leaf Skeletonizer Moth
Tuis, Cartago, Costa Rica

Pyralidae: Pyralinae
Mapeta xanthomelas • Yellow and Black Snout Moth
Playa Los Angeles, Magdalena, Colombia

Crambidae: Spilomelinae
Asturodes fimbriauralis • Gold-fringed Snout Moth
Montagne de Kaw, Roura, French Guiana

Crambidae: Spilomelinae
Diaphania cf. costata • Costate Diaphania
Montagne de Kaw, Roura, French Guiana

Crambidae: Spilomelinae
Conchylodes platinalis • Snout Moth
Tuis, Cartago, Costa Rica

Crambidae: Spilomelinae
Diaphania esmeraldus • Emerald Diaphania
Montagne de Kaw, Roura, French Guiana

Crambidae: Spilomelinae
Diacme cf. mopsalis • Snout Moth
Playa Los Angeles, Magdalena, Colombia

Crambidae: Spilomelinae
Diaphania nitidalis • Shining Diaphania
Tuis, Cartago, Costa Rica

Crambidae: Spilomelinae
Glyphodes rubrocinctalis • Red-belted Snout Moth
Tuis, Cartago, Costa Rica

Crambidae: Spilomelinae
Pycnarmon sp. • Spilomeline Snout Moth
Playa Los Angeles, Magdalena, Colombia

Crambidae: Spilomelinae
Omiodes fulvicauda • Spilomeline Snout Moth
Tuis, Cartago, Costa Rica

Crambidae: Musotiminae
≈ Undulambia sp. • Musotimine Moth
Tuis, Cartago, Costa Rica

Crambidae: Spilomelinae
Omiodes pandaralis • Spilomeline Snout Moth
Tuis, Cartago, Costa Rica

Crambidae: Acentropinae
Petrophila sp. • Rock-loving Crambid
Tuis, Cartago, Costa Rica

Crambidae: Pyraustinae
Desmia sp. • Leaf-folder Moth
Playa Los Angeles, Magdalena, Colombia

Crambidae: Pyraustinae
Syngamia florella • Little Flower Moth
Manaus, Amazonas, Brazil

Crambidae: Pyraustinae
Maruca vitrata testulalis • Bean Pod Borer
Montagne de Kaw, Roura, French Guiana

Crambidae: Midilinae
Midila sp. • Midiline Snout Moth
Tuis, Cartago, Costa Rica

Crambidae: Pyraustinae
Syllepis hortalis • Snout Moth
Parque Nacional Pico Bonito, lodge, Atlántida, Honduras

Crambidae: Midilinae
Hositea gynaecia • Midiline Snout Moth
Montagne de Kaw, Roura, French Guiana

Crambidae: Midilinae
Eupastranaia tumidifrons • Midiline Snout Moth
Montagne de Kaw, Roura, French Guiana

Mimallonidae: Lacosominae
Tolypida amaryllis • Sack-bearer Moth
Montagne de Kaw, Roura, French Guiana

Crambidae: Crambinae
Argyria cf. croceivittella • Snout Moth
Rewa village, Upper Takutu-Upper Essequibo, Guyana

Mimallonidae: Incertae sedis
Roelofa cf. hegewischi • Sack-bearer Moth
Mountain Pine Ridge Forest Reserve, Belize

Mimallonidae: Cicinninae
Psychocampa concolor • Same-colored Psychocampa
Colakreek, Para, Suriname

Apatelodidae: Apatelodinae
Apatelodes pandariodes • American Silkworm Moth
Parque Nacional Pico Bonito, lodge, Atlántida, Honduras

Apatelodidae: Apatelodinae
Apatelodes nr. pithala • American Silkworm Moth
Berg en Dal, Brokopondo, Suriname

Apatelodidae: Apatelodinae
Quentalia surynorta • American Silkworm Moth
Tuis, Cartago, Costa Rica

Apatelodidae: Apatelodinae
Ephoria sp. • American Silkworm Moth
Playa Los Angeles, Magdalena, Colombia

Bombycidae: Bombycinae
Anticla antica • Silk Moth
Gamboa, Colón, Panama

Apatelodidae: Apatelodinae
Olceclostera nigripuncta • American Silkworm Moth
Gamboa, Colón, Panama

Saturniidae: Arsenurinae
Rhescyntis hippodamia • Hemileucine Giant Silkmoth
Tuis, Cartago, Costa Rica

Saturniidae: Ceratocampinae
Adeloneivaia jason • Jason's Royal Moth
Tuis, Cartago, Costa Rica

Saturniidae: Oxyteninae
Oxytenis sp. • Tropical American Silkworm Moth
Gamboa, Colón, Panama

Saturniidae: Ceratocampinae
Orthorene cf. verana • Royal Moth
Mountain Pine Ridge Forest Reserve, Belize

Saturniidae: Oxyteninae
Therinia lactucina • Tropical American Silkworm Moth
Along the Rupununi River, Guyana

Saturniidae: Oxyteninae
Oxytenis cf. mirabilis • Tropical American Silkworm Moth
Gamboa, Colón, Panama

Saturniidae: Hemileucinae
Gamelia rindgei • Rindge's Gamelia
Montagne de Kaw, Roura, French Guiana

Saturniidae: Hemileucinae
Hylesia sp. • Hemileucine Giant Silkmoth
Berg en Dal, Brokopondo, Suriname

Saturniidae: Hemileucinae
Pseudodirphia imperialis • Imperial Hemileucine Moth
Gamboa, Colón, Panama

Saturniidae: Hemileucinae
Leucanella acutissima • Hemileucine Giant Silkmoth
Mountain Pine Ridge Forest Reserve, Belize

Saturniidae: Hemileucinae
Automeris nr. curvilinea • Eyed Silkmoth
Berg en Dal, Brokopondo, Suriname

Saturniidae: Hemileucinae
Periphoba arcaei • Hemileucine Giant Silkmoth
Tuis, Cartago, Costa Rica

Saturniidae: Hemileucinae
Automeris postalbida • Eyed Silkmoth
Gamboa, Colón, Panama

Saturniidae: Hemileucinae

Pseudodirphia menander • Hemileucine Giant Silkmoth

Tuis, Cartago, Costa Rica

Sphingidae: Macroglossinae

Nyceryx riscus • Macroglossine Sphinx Moth

Reserva Particular do Cristalino, Mato Grosso, Brazil

Sphingidae: Sphinginae

Manduca rustica • Rustic Sphinx

Montagne de Kaw, Roura, French Guiana

Sphingidae: Macroglossinae

Erinnyis crameri • Cramer's Sphinx

Playa Los Angeles, Magdalena, Colombia

Sphingidae: Sphinginae

Adhemarius eurysthenes • Broad-breasted Sphinx

Montagne de Kaw, Roura, French Guiana

Sphingidae: Macroglossinae

Pachylia darceta • Macroglossine Sphinx Moth

Montagne de Kaw, Roura, French Guiana

Sphingidae: Macroglossinae
Pachylia ficus • Fig Sphinx
Sansón, Darién, Panama

Sphingidae: Sphinginae
Lintneria sp. • Sphinx Moth
Parque Nacional Pico Bonito, lodge, Atlántida, Honduras

Sphingidae: Macroglossinae
Callionima nomius • Fan-tailed Bark Moth
Valle de Antón, Coclé, Panama

Sphingidae: Macroglossinae
Aellopos ceculus• Ceculus Sphinx
Valle de Antón, Coclé, Panama

Sphingidae: Macroglossinae
Madoryx plutonius • Dusky Madoryx
Parque Nacional Pico Bonito, lodge, Atlántida, Honduras

Sphingidae: Macroglossinae
Eumorpha capronnieri • Macroglossine Sphinx Moth
Montagne de Kaw, Roura, French Guiana

Sphingidae: Macroglossinae
Eumorpha obliquus • Oblique Eumorpha
Montagne de Kaw, Roura, French Guiana

Sphingidae: Macroglossinae
Xylophanes belti • Belt's Spinx
Tuis, Cartago, Costa Rica

Sphingidae: Macroglossinae
Eumorpha phorbas • Pasture Eumorpha
Parque Nacional Pico Bonito, lodge, Atlántida, Honduras

Sphingidae: Macroglossinae
Xylophanes ceratomioides • Macroglossine Sphinx
Tuis, Cartago, Costa Rica

Sphingidae: Macroglossinae
Eumorpha triangulum • Triangular Eumorpha
Tuis, Cartago, Costa Rica

Sphingidae: Macroglossinae
Xylophanes chiron • Chiron sphinx
Tuis, Cartago, Costa Rica

Sphingidae: Macroglossinae
Xylophanes cyrene • Cyrene Sphinx
Montagne de Kaw, Roura, French Guiana

Lasiocampidae: Poecilocampinae
Euglyphis sp. • Lappet Moth
Berg en Dal, Brokopondo, Suriname

Sphingidae: Macroglossinae
Xylophanes loelia •Loelia Sphinx
Berg en Dal, Brokopondo, Suriname

Lasiocampidae: Poecilocampinae
Titya cf. cinerascens • Poodle Moth
Montagne de Kaw, Roura, French Guiana

Lasiocampidae: Poecilocampinae
Euglyphis sp. • Lappet Moth
Berg en Dal, Brokopondo, Suriname

Uraniidae: Uraniinae
Urania leilus • Urania Moth
Berg en Dal, Brokopondo, Suriname

Geometridae: Ennominae
Epimecis nr. matronaria • Ennomine Geometer
Leona Vicario, Quintana Roo, Mexico

Geometridae: Ennominae
Iridopsis nr. larvaria • Ennomine Geometer
Mountain Pine Ridge Forest Reserve, Belize

Geometridae: Ennominae
Erastria decrepitaria • Ennomine Geometer
Long Bay, Portland parish, Jamaica

Geometridae: Ennominae
Iridopsis validaria • Ennomine Geometer
Tuis, Cartago, Costa Rica

Geometridae: Ennominae
Nematocampa completa • Filament-bearer
Boquete, Chiriquí, Panama

Geometridae: Ennominae
Leuciris fimbriaria • Fringed Leuciris
Playa Los Angeles, Magdalena, Colombia

Geometridae: Ennominae
Opisthoxia bolivari • Bolívar's Ennomine Moth
Tuis, Cartago, Costa Rica

Geometridae: Ennominae
Macaria gambarina • Macaria Geometer Moth
Tuis, Cartago, Costa Rica

Geometridae: Ennominae
Sphacelodes vulneraria • Wounded Geometer
Leona Vicario, Quintana Roo, Mexico

Geometridae: Ennominae
Paragonia cruraria • Pale Beauty
Montagne de Kaw, Roura, French Guiana

Geometridae: Ennominae
Cyphoedma transvolutata • Ennomine Geometer
Tuis, Cartago, Costa Rica

Geometridae: Ennominae
Tmetomorpha bitias • Ennomine Geometer
Tuis, Cartago, Costa Rica

Geometridae: Ennominae
Sericoptera chiffa • Ennomine Geometer
Tuis, Cartago, Costa Rica

Geometridae: Geometrinae
Chloropteryx opalaria • Mottled Opal
Tuis, Cartago, Costa Rica

Geometridae: Ennominae
Simena luctifera • Ennomine Moth
Tuis, Cartago, Costa Rica

Geometridae: Geometrinae
Oospila circumsignata • Edge-marked Emerald
Montagne de Kaw, Roura, French Guiana

Geometridae: Geometrinae
Chloropteryx sp. • Emerald
Crooked Tree Preserve, Belize district, Belize

Geometridae: Geometrinae
Oospila venezuelata • Venezuelan Emerald
Tuis, Cartago, Costa Rica

Geometridae: Geometrinae
Oospila albicoma • White-haired Emerald
Tuis, Cartago, Costa Rica

Geometridae: Geometrinae
Rhodochlora brunneipalpis • Rosy-green Emerald
Tuis, Cartago, Costa Rica

Geometridae: Geometrinae
Nemoria astraea • Starry Emerald
Tuis, Cartago, Costa Rica

Geometridae: Geometrinae
Synchlora astraeoides • Emerald
Tuis, Cartago, Costa Rica

Geometridae: Geometrinae
Nemoria tickelli • Tickell's Emerald
Tuis, Cartago, Costa Rica

Geometridae: Geometrinae
Synchlora ephippiaria • Saddled Emerald
Presqu'île de la Caravelle, Martinique

Geometridae: Geometrinae
Synchlora gerularia • Emerald
Berg en Dal, Brokopondo, Suriname

Geometridae: Geometrinae
Nemoria vermiculata • Emerald Moth
Boquete, Chiriquí, Panama

Geometridae: Geometrinae
Synchlora pulchrifimbria • Beautiful Fringed Emerald
Parque Nacional Pico Bonito, lodge, Atlántida, Honduras

Geometridae: Geometrinae
Hydata busa • Emerald
Tuis, Cartago, Costa Rica

Geometridae: Geometrinae
Oospila decorata. • Decorated Emerald
Montagne de Kaw, Roura, French Guiana

Geometridae: Sterrhinae
Hemipterodes rotundata • Rotund Sterrhine Geometer
Gamboa, Colón, Panama

Geometridae: Sterrhinae
Idaea subfervens • Sterrhine Geometer
Montagne de Kaw, Roura, French Guiana

Geometridae: Larentiinae
Eois paraviolascens • Larentiine Geometer
Nanegalito, Tandayapa Valley, Ecuador

Geometridae: Sterrhinae
Leptostales sp. • Sterrhine Geometer
Berg en Dal, Brokopondo, Suriname

Geometridae: Larentiinae
Eois fasciata • Banded Eois
Berg en Dal, Brokopondo, Suriname

Geometridae: Sterrhinae
Prasinochrysa quadriplaga • Sterrhine Geometer
Berg en Dal, Brokopondo, Suriname

Geometridae: Larentiinae
Eois sp. • Larentiine Geometer
Tuis, Cartago, Costa Rica

Geometridae: Larentiinae
Eubaphe cf. unicolor • Larentiine Geometer
Playa Los Angeles, Magdalena, Colombia

Notodontidae: Nystaleinae
Strophocerus cossoides • Prominent Moth
Montagne de Kaw, Roura, French Guiana

Geometridae: Larentiinae
Oligopleura aulaeata • Larentiine Geometer
Tuis, Cartago, Costa Rica

Notodontidae: Nystaleinae
Calledema sp. • Prominent Moth
Montagne de Kaw, Roura, French Guiana

Notodontidae: Rifargiriinae
Magava multilinea • Many-lined Prominent
Montagne de Kaw, Roura, French Guiana

Notodontidae: Nystaleinae
Pentobesa cf. xylinoides • Twig Moth
Tuis, Cartago, Costa Rica

Notodontidae: Heterocampinae
Disphragis aemula • Owlet Moth
Berg en Dal, Brokopondo, Suriname

Notodontidae: Heterocampinae
Malocampa punctata • Heterocampine Prominent
Gamboa, Colón, Panama

Notodontidae: Heterocampinae
Eragisa antarorum • Heterocampine Prominent
Montagne de Kaw, Roura, French Guiana

Notodontidae: Heterocampinae
Rosema aethra • Heterocampine Prominent
Sansón, Darién, Panama

Notodontidae: Heterocampinae
Farigia sp. • Heterocampine Prominent
Gamboa, Colón, Panama

Notodontidae: Heterocampinae
Rosema thestia • Heterocampine Prominent
Tuis, Cartago, Costa Rica

Erebidae: Anobinae
Anoba pohli • Anobine Moth
Leona Vicario, Quintana Roo, Mexico

Erebidae: Calpinae
Dolichosomastis sp. • Fruit-Piercing Moth
Berg en Dal, Brokopondo, Suriname

Erebidae: Calpinae
Ostha nr. rama • Fruit-Piercing Moth
Tuis, Cartago, Costa Rica

Erebidae: Calpinae
Gonodonta sp. • Fruit-Piercing Moth
Comarca Indígena Emberá-Wounaan, Panama

Erebidae: Calpinae
Gorgone augusta • Augusta Gorgone
Gamboa, Colón, Panama

Erebidae: Calpinae
Gorgone sp. • Fruit-piercing Moth
Berg en Dal, Brokopondo, Suriname

Erebidae: Eulepidotinae
Antiblemma trogocycla • Eulepidotine Moth
Tuis, Cartago, Costa Rica

Erebidae: Erebinae
Anticarsia sp. • Erebine Moth
Along the Rio Negro, Amazonas state, Brazil

Erebidae: Eulepidotinae
Azeta sp. • Eulepidotine Moth
Along the Rupununi River, Guyana

Erebidae: Erebinae
Ctypansa cf. obtusa • Erebine Moth
Montagne de Kaw, Roura, French Guiana

Erebidae: Eulepidotinae
Epitausa dilina • Two-lined Epitausa
Tuis, Cartago, Costa Rica

Erebidae: Erebinae
Focillini • Erebine Moth
Along the Teles Pires River, Mato Grosso, Brazil

Erebidae: Erebinae
Ocalaria sp. • Erebine Moth
Valle de Antón, Coclé, Panama

Erebidae: Arctiinae
Agaraea cf. schausi • Tiger Moth
Tuis, Cartago, Costa Rica

Erebidae: Herminiinae
Mastixis galealis • Litter Moth
Tuis, Cartago, Costa Rica

Erebidae: Arctiinae
Dysschema magdala • Tiger Moth
Tuis, Cartago, Costa Rica

Erebidae: Arctiinae
Cratoplastis diluta • Firefly Mimic Tiger Moth
Gamboa, Colón, Panama

Erebidae: Arctiinae
Dysschema cerialis • Tiger Moth
Tuis, Cartago, Costa Rica

Erebidae: Arctiinae
Hyperandra diminuta • Diminished Tiger Moth
Berg en Dal, Brokopondo, Suriname

Erebidae: Arctiinae
Utetheisa ornatrix • Ornate Bella Moth
Crooked Tree Preserve, Belize district, Belize

Erebidae: Arctiinae
Lophocampa annulosa • Spotted Tussock Moth
Leona Vicario, Quintana Roo, Mexico

Erebidae: Arctiinae
Phaloe cubana • Cuban Tiger Moth
Leona Vicario, Quintana Roo, Mexico

Erebidae: Arctiinae
Correbia lycoides • Lycid Mimic Tiger Moth
Rewa village, Upper Takutu-Upper Essequibo, Guyana

Erebidae: Arctiinae
Areva trigemmis • Lichen Moth
Valle de Antón, Coclé, Panama

Erebidae: Arctiinae
Clemensia leopardina • Leopard Lichen Moth
Tuis, Cartago, Costa Rica

Erebidae: Arctiinae
Rhipha chionoplaga • Arctiine Moth
Mountain Pine Ridge Forest Reserve, Belize

Erebidae: Arctiinae
Rhabdatomis cf. laudamia • Lichen Moth
Leona Vicario, Quintana Roo, Mexico

Erebidae: Arctiinae
Bertholdia specularis • Arctiine Moth
Tuis, Cartago, Costa Rica

Erebidae: Arctiinae
Tuina cingulata • Lichen Moth
Tuis, Cartago, Costa Rica

Erebidae: Arctiinae
Trichromia cf. cotes • Arctiine Moth
Crooked Tree Preserve, Belize district, Belize

Erebidae: Arctiinae
Trichromia cf. cotes • Arctiine Moth
Leona Vicario, Quintana Roo, Mexico

Erebidae: Arctiinae
Trichromia cf. cardinalis • Cardinal Tiger Moth
Berg en Dal, Brokopondo, Suriname

Erebidae: Arctiinae
Trichromia flavoroseus (dark form) • Arctiine Moth
Leona Vicario, Quintana Roo, Mexico

Erebidae: Arctiinae
Trichromia postflavida • Arctiine Moth
Montagne de Kaw, Roura, French Guiana

Erebidae: Arctiinae
Theages decora • Arctiine Moth
Tuis, Cartago, Costa Rica

Erebidae: Arctiinae
Eucereon aroa • Arctiine Moth
Tuis, Cartago, Costa Rica

Erebidae: Arctiinae
Eucereon confinis • Arctiine Moth
Boquete, Chiriquí province, Panama

Erebidae: Arctiinae
Eucereon sp. • Arctiine Moth
Playa Los Angeles, Magdalena, Colombia

Erebidae: Arctiinae
Eucereon cf. latifascia • Arctiine Moth
Montagne de Kaw, Roura, French Guiana

Erebidae: Arctiinae
Eucereon tigrata• Arctiine Moth
Tuis, Cartago, Costa Rica

Erebidae: Arctiinae
Eucereon cf. costulata • Arctiine Moth
Boquete, Chiriquí, Panama

Erebidae: Arctiinae
Phaeomolis lineatus • Lined Tiger Moth
Parque Nacional Pico Bonito, lodge, Atlántida, Honduras

Erebidae: Arctiinae
Xenosoma geometrina • Arctiine Moth
Tuis, Cartago, Costa Rica

Erebidae: Arctiinae
Mydromera notochloris • Green-marked Mydromera
Tuis, Cartago, Costa Rica

Erebidae: Arctiinae
Macrocneme chrysitis • Southern Cyan Tiger Moth
Parque Nacional Pico Bonito, lodge, Atlántida, Honduras

Erebidae: Arctiinae
Syntomeida cf. melanthus • Wasp Moth
Along the Rupununi River, Guyana

Erebidae: Arctiinae
Calonotos helymus • Arctiine Moth
Montagne de Kaw, Roura, French Guiana

Erebidae: Arctiinae
Homoeocera stictosoma • Wasp Moth
Gamboa, Colón, Panama

Erebidae: Arctiinae
Cosmosoma nr. achemon • Wasp Moth
Playa Los Angeles, Magdalena, Colombia

Erebidae: Arctiinae
Cosmosoma stibostictum • Wasp Moth
Parque Nacional Pico Bonito, lodge, Atlántida, Honduras

Erebidae: Arctiinae
Cosmosoma festiva • Festive Wasp Moth
Tuis, Cartago, Costa Rica

Erebidae: Arctiinae
Leucotmemis (torrida group) sp. • Wasp Moth
Reserva Particular do Cristalino, Mato Grosso, Brazil

Erebidae: Arctiinae
Cosmosoma jeovana • Wasp Moth
Boquete, Chiriquí province, Panama

Erebidae: Arctiinae
Loxophlebia davisi • Davis's Wasp Moth
Playa Los Angeles, Magdalena, Colombia

Erebidae: Arctiinae
Evius albicoxae • Arctiine Moth
Berg en Dal, Brokopondo, Suriname

Erebidae: Arctiinae
Amaxia juvenis • Arctiine Moth
Leona Vicario, Quintana Roo, Mexico

Erebidae: Arctiinae
Parevia nr. metachryseis • Arctiine Moth
Montagne de Kaw, Roura, French Guiana

Erebidae: Arctiinae
Amaxia pandama • Arctiine Moth
Berg en Dal, Brokopondo, Suriname

Erebidae: Arctiinae
Amaxia bella • Pretty Amaxia
Montagne de Kaw, Roura, French Guiana

Erebidae: Arctiinae
Idalus dares • Tiger Moth
Parque Nacional Pico Bonito, lodge, Atlántida, Honduras

Erebidae: Arctiinae
Idalus crinis • Clown Face Tiger Moth
Tuis, Cartago, Costa Rica

Erebidae: Arctiinae
Amastus aconia rumina • Arctiine Moth
Boquete, Chiriquí province, Panama

Erebidae: Arctiinae
Idalus occidentalis • Western Idalus
Soberanía National Park, Colón, Panama

Erebidae: Arctiinae
Coiffaitarctia steniptera • Arctiine Moth
Montagne de Kaw, Roura, French Guiana

Erebidae: Arctiinae
Turuptiana obliqua • Oblique Turuptiana
Tuis, Cartago, Costa Rica

Erebidae: Arctiinae
Himerarctia docis • Arctiine Moth
Montagne de Kaw, Roura, French Guiana

Erebidae: Arctiinae
Ormetica ataenia • Arctiine Moth
Tuis, Cartago, Costa Rica

Erebidae: Arctiinae
Isanthrene monticola • Mountain-dwelling Arctiine
Boquete, Chiriquí, Panama

Erebidae: Arctiinae
Ormetica cf. *orbona* • Arctiine Moth
Parque Nacional Pico Bonito, lodge, Atlántida, Honduras

Erebidae: Erebinae
Letis buteo • Letis Moth
Montagne de Kaw, Roura, French Guiana

Erebidae: Arctiinae
Ormetica contraria • Contrary Tiger Moth
Montagne de Kaw, Roura, French Guiana

Erebidae: Erebinae
Letis sp. • Letis Moth
Tuis, Cartago, Costa Rica

Erebidae: Erebinae
Feigeria herilia • Letis Moth
Montagne de Kaw, Roura, French Guiana

Erebidae: Erebinae
Mocis sp. • Erebine Moth
Along the Rupununi River, Guyana

Erebidae: Erebinae
Hemeroblemma dolosa • Erebine Moth
Tuis, Cartago, Costa Rica

Nolidae: Eligminae
Iscadia furcifera • Nolid Moth
Mountain Pine Ridge Reserve, Cayo, Belize

Erebidae: Erebinae
Metria bruma • Winter Metria
Tuis, Cartago, Costa Rica

Nolidae: Nolinae
Meganola polyodonta • Scribble Moth
Playa Los Angeles, Magdalena, Colombia

Noctuidae: Cydosiinae
Cydosia curvinella • Small-curved Owlet Moth
Lago Bayano, Panamá province, Panama

Noctuidae: Dyopsinae
Sosxetra grata • Pleasing Sosxetra
Parque Nacional Pico Bonito, lodge, Atlántida, Honduras

Noctuidae: Dyopsinae
Ceroctena amynta • Emeraldine
Tuis, Cartago, Costa Rica

Noctuidae: Noctuinae
Bryolymnia bicon • Dart Moth
Boquete, Chiriquí, Panama

Noctuidae: Lymantriinae
Caviria cf. vinasia • Tussock Moth
Mountain Pine Ridge Forest Reserve, Belize

Noctuidae: Eriopinae
Callopistria meres • Fern Moth
Playa Los Angeles, Magdalena, Colombia

Papilionidae: Papilioninae
Battus polydamas • Polydamus Swallowtail
12 km south of Filadelfia, Boquerón, Paraguay

Papilionidae: Papilioninae
Papilio thoas • Thoas Swallowtail
Lago Bayano, Panamá province, Panama

Papilionidae: Papilioninae
Papilio polyxenes • Costa Rican Black Swallowtail
Monteverde, Puntarenas, Costa Rica

Papilionidae: Papilioninae
Parides iphidamas • Iphidamas Cattleheart
Playa Los Angeles, Magdalena, Colombia

Papilionidae: Papilioninae
Papilio androgeus • Androgeus Swallowtail ♂
Along the Teles Pires River, Mato Grosso, Brazil

Papilionidae: Papilioninae
Parides panares • Wedge-spotted Cattleheart
Suchitoto, Cuscatlán, El Salvador

Pieridae: Coliadinae
Anteos maerula • Yellow Angled-Sulphur
Amatlán de los Reyes, Veracruz, Mexico

Pieridae: Coliadinae
Phoebis argante • Apricot Sulphur
Parque Nacional Tortuguero, Limón, Costa Rica

Pieridae: Coliadinae
Eurema cf. daira • Barred Yellow
Naranjo, Alajuela province, Costa Rica

Pieridae: Coliadinae
Phoebis philea • Orange-barred Sulphur
Cacaxtla, Tlaxcala, Mexico

Pieridae: Coliadinae
Eurema sp. • Grass Yellow
Cabo Rojo, Pedernales, Dominican Republic

Pieridae: Coliadinae
Phoebis sennae marcellina • Cloudless Sulphur
Playa Los Angeles, Magdalena, Colombia

Pieridae: Dismorphiinae
Dismorphia crisia lubina • Crisia Mimic-White
Boquete, Chiriquí, Panama

Pieridae: Pierinae
Melete lycimnia • Common Melwhite
Gamboa, Colón, Panama

Pieridae: Dismorphiinae
Dismorphia theucharila • Clear-wing Mimic White
Selva Negra, Matagalpa, Nicaragua

Pieridae: Pierinae
Tatochila mercedis vanvolxemii • Mercedis White
Puerto Madryn, Chubut, Argentina

Pieridae: Pierinae
Leptophobia aripa • Green-eyed White
Selva Negra, Matagalpa, Nicaragua

Pieridae: Pierinae
Tatochila mercedis sterodice • Mercedis White
Puerto Madryn, Chubut, Argentina

Pieridae: Pierinae
Ascia monuste • Great Southern White
Filadelfia, Boquerón, Paraguay

Lycaenidae: Theclinae
Arawacus separata • Separated Stripestreak
Along the Rupununi River, Guyana

Lycaenidae: Theclinae
Arawacus togarna • Togama Stripestreak
Santo Domingo de los Colorados, Pichincha, Ecuador

Lycaenidae: Theclinae
Megathecla cf. cupentius • Hairstreak
Berg en Dal, Brokopondo, Suriname

Lycaenidae: Theclinae
Arawacus aetolus • Aetolus Stripestreak
Berg en Dal, Brokopondo, Suriname

Lycaenidae: Theclinae
Strymon sp. • Hairstreak
Berg en Dal, Brokopondo, Suriname

Lycaenidae: Theclinae
Strymon astiocha • Scrub Hairstreak
Manaus, Amazonas, Brazil

Lycaenidae: Theclinae
Megathecla cf. cupentius • Hairstreak
Berg en Dal, Brokopondo, Suriname

Lycaenidae: Theclinae
Tmolus echion • Red-spotted Hairstreak
Along the Rupununi River, Guyana

Lycaenidae: Theclinae
Calycopis isobeon • Dusky-blue Groundstreak
Gamboa, Colón, Panama

Lycaenidae: Theclinae
Strymon megarus • Scrub Hairstreak
Comarca Indígena Emberá-Wounaan, Panama

Lycaenidae: Theclinae
Calycopis sp. Malta species group • Groundstreak
Colakreek, Para, Suriname

Lycaenidae: Theclinae
Calycopis sp. • Groundstreak
Montsinéry, Montsinéry-Tonnegrande, French Guiana

Lycaenidae: Polyommatinae
Hemiargus hanno • Hanno Blue
Novo Airão, Amazonas, Brazil

Lycaenidae: Theclinae
Calycopis cf. atnius • Atnius Groundstreak
Along the Rupununi River, Guyana

Lycaenidae: Polyommatinae
Cupido comyntas • Eastern Tailed Blue
Mountain Pine Ridge Forest Reserve, Belize

Lycaenidae: Theclinae
Eumaeus toxea • Mexican Cycadian
Fortín de las Flores, Veracruz, Mexico

Lycaenidae: Polyommatinae
Leptotes cassius • Cassius Blue
Filadelfia, Boquerón, Paraguay

Riodinidae: Riodininae
Lasaia agesilas • Black-patched Metalmark
Along the Teles Pires River, Mato Grosso, Brazil

Riodinidae: Riodininae
Mesosemia lamachus • Lamachus Eyemark
Leona Vicario, Quintana Roo, Mexico

Riodinidae: Riodininae
Eurybia lycisca • Blue-winged Euyrbia
Parque Natural Metropolitano, Panamá, Panama

Riodinidae: Riodininae
Mesosemia telegone • Telegone Eyemark
Playa Los Angeles, Magdalena, Colombia

Riodinidae: Riodininae
Mesosemia machaera • Dagger Eyemark
Along the Rio Negro, Amazonas, Brazil

Riodinidae: Riodininae
Theope nycteis • Eyed Theope
Along the Rewa River, Guyana

Riodinidae: Riodininae
Amarynthis meneria • Meneria Metalmark
Reserva Particular do Cristalino, Mato Grosso, Brazil

Riodinidae: Riodininae
Perophthalma tullius • Tullius Eyemark
Altos del Maria, Coclé, Panama

Riodinidae: Riodininae
Riodina lysippus • Scintillant Metalmark
Berg en Dal, Brokopondo, Suriname

Riodinidae: Riodininae
Stalachtis nr. calliope • Scintillant Metalmark
Along the Rewa River, Guyana

Riodinidae: Riodininae
Detrivora sp. • Scintillant Metalmark
Berg en Dal, Brokopondo, Suriname

Riodinidae: Riodininae
Calospila lucianus • Lucianus Metalmark
Berg en Dal, Brokopondo, Suriname

Riodinidae: Riodininae
Calospila emylius • Emylius Metalmark
Novo Airão, Amazonas, Brazil

Riodinidae: Riodininae
Aricoris propitia • Propitious Metalmark
Novo Airão, Amazonas, Brazil

Riodinidae: Riodininae
Setabis epitus • Epitus Metalmark
Montsinéry, Montsinéry-Tonnegrande, French Guiana

Riodinidae: Riodininae
Calephelis sp. • Scintillant Metalmark
Selva Negra, Matagalpa, Nicaragua

Riodinidae: Riodininae
Anteros formosus • Beautiful Jewelmark
Parque Nacional Pico Bonito, Atlántida, Honduras

Riodinidae: Riodininae
Calephelis sp. • Scintillant Metalmark
Playa Los Angeles, Magdalena, Colombia

Riodinidae: Riodininae
Calephelis fulmen • Lightning Metalmark
Leona Vicario, Quintana Roo, Mexico

Riodinidae: Riodininae
Melanis sp. • Pixie
Along the Teles Pires River, Mato Grosso, Brazil

Riodinidae: Riodininae
Nymphidium ascolia • Creamy Metalmark
Gamboa, Colón, Panama

Nymphalidae: Heliconiinae
Actinote parapheles • Parapheles Heliconian
Colonia del Sacramento, Colonia, Uruguay

Riodinidae: Riodininae
Melanis smithiae • Smith's Pixie
Novo Airão, Amazonas, Brazil

Nymphalidae: Heliconiinae
Agraulis vanillae • Gulf Fritillary
Playa Los Angeles, Magdalena, Colombia

Nymphalidae: Heliconiinae
Euptoieta hegesia • Mexican Fritillary
Amatlán, Vera Cruz, Mexico

Nymphalidae: Heliconiinae
Dryadula phaetusa • Banded Orange Heliconian
Novo Airão, Amazonas, Brazil

Nymphalidae: Heliconiinae
Dione juno • Juno Longwing
Monteverde, Puntarenas, Costa Rica

Nymphalidae: Heliconiinae
Heliconius charithonia • Zebra Longwing
Leona Vicario, Quintana Roo, Mexico

Nymphalidae: Heliconiinae
Dione moneta • Mexican Silverspot
Zitácuaro, Michoacán, Mexico

Nymphalidae: Heliconiinae
Heliconius clysonymus • Clysonymus longwing
Monteverde, Puntarenas, Costa Rica

Nymphalidae: Heliconiinae
Heliconius erato cyrbia • Red Postman
Santo Domingo de los Colorados, Pichincha, Ecuador

Nymphalidae: Heliconiinae
Heliconius demeter • Demeter Longwing
Novo Airão, Amazonas, Brazil

Nymphalidae: Heliconiinae
Heliconius erato hydara • Crimson-patched Longwing
Playa Los Angeles, Magdalena, Colombia

Nymphalidae: Heliconiinae
Heliconius sara • Sara Longwing
Asunción, Cusco Department, Peru

Nymphalidae: Heliconiinae
Heliconius erato petiverana • Crimson-patched Longwing
Leona Vicario, Quintana Roo, Mexico

Nymphalidae: Heliconiinae
Heliconius ethilla • Ethilia Longwing
Rewa village, Upper Takutu-Upper Essequibo, Guyana

Nymphalidae: Heliconiinae
Heliconius hecale • Hecale Longwing
Monteverde, Puntarenas, Costa Rica

Nymphalidae: Nymphalinae
Anthanassa drusilla • Orange-patched Crescent
Santo Domingo de los Colorados, Pichincha, Ecuador

Nymphalidae: Heliconiinae
Neruda aoede • Aoede Longwing
Berg en Dal, Brokopondo, Suriname

Nymphalidae: Nymphalinae
Anthanassa ardys • Ardys Crescent ♀
Valle de Antón, Coclé, Panama

Nymphalidae: Heliconiinae
Philaethria dido • Green Longwing
Shintuya Hot Springs, Madre de Dios, Peru

Nymphalidae: Nymphalinae
Chlosyne janais • Crimson Patch
Amatlán de los Reyes, Veracruz, Mexico

Nymphalidae: Nymphalinae
Chlosyne lacinia • Bordered Patch
Lago Bayano, Panamá province, Panama

Nymphalidae: Nymphalinae
Ortilia velica • Velica Crescent
Punta Gorda, Nueva Palmira, Colonia, Uruguay

Nymphalidae: Nymphalinae
Eresia clio • Common Crescent
Gamboa, Colón, Panama

Nymphalidae: Nymphalinae
Tegosa claudina • Claudina Crescent
Punta Gorda, Nueva Palmira, Colonia, Uruguay

Nymphalidae: Nymphalinae
Ortilia ithra • Ithra Crescent
Colonia del Sacramento, Colonia, Uruguay

Nymphalidae: Nymphalinae
Vanessa carye • South American Painted Lady
Puerto Madryn, Chubut, Argentina

Nymphalidae: Nymphalinae
Anartia amathea • Scarlett Peacock
Santo Domingo de los Colorados, Pichincha, Ecuador

Nymphalidae: Nymphalinae
Historis odius • Orion Cecropian
Parque Natural Metropolitano, Panamá, Panama

Nymphalidae: Nymphalinae
Anartia jatrophae • White Peacock
Monteverde, Puntarenas, Costa Rica

Nymphalidae: Nymphalinae
Siproeta epaphus • Rusty-tipped Page
Santa Elena, Puntarenas, Costa Rica

Nymphalidae: Nymphalinae
Anartia fatima • Banded Peacock
Selva Negra, Matagalpa, Nicaragua

Nymphalidae: Nymphalinae
Siproeta epaphus • Rusty-tipped Page
Santa Elena, Puntarenas, Costa Rica

Nymphalidae: Nymphalinae
Siproeta stelenes • Malachite
Long Bay, Portland parish, Jamaica

Nymphalidae: Apaturinae
Doxocopa laurentia • Turquoise Emperor
Colonia del Sacramento, Colonia, Uruguay

Nymphalidae: Nymphalinae
Colobura dirce dirce • Zebra Mosaic
Parque Nacional Pico Bonito, Atlántida, Honduras

Nymphalidae: Cyrestinae
Marpesia berania • Orange Daggerwing
Tuis, Cartago, Costa Rica

Nymphalidae: Nymphalinae
Junonia evarete • Tropical Buckeye
Berg en Dal, Brokopondo, Suriname

Nymphalidae: Cyrestinae
Marpesia chiron • Many-banded Daggerwing
Reserva Particular do Cristalino, Mato Grosso, Brazil

Nymphalidae: Cyrestinae
Marpesia orsilochus • Orsilochus Daggerwing
Reserva Particular do Cristalino, Mato Grosso, Brazil

Nymphalidae: Biblidinae
Diaethria pandama • Orange-striped Eighty-eight
Selva Negra, Matagalpa, Nicaragua

Nymphalidae: Cyrestinae
Marpesia petreus • Stone Daggerwing
Along the Teles Pires River, Mato Grosso, Brazil

Nymphalidae: Biblidinae
Paulogramma pyracmon • Pyracmon Eighty-eight
Atalaya, Cusco Department, Peru

Nymphalidae: Biblidinae
Diaethria candrena • Number Eighty
Along Arroyo de las Víboras, Colonia, Uruguay

Nymphalidae: Biblidinae
Hamadryas amphichloe diasia • Caribbean Cracker
Cabo Rojo, Pedernales, Dominican Republic

Nymphalidae: Biblidinae
Hamadryas iphthime • Iphthime Cracker
Parque Nacional Pico Bonito, Atlántida, Honduras

Nymphalidae: Biblidinae
Mestra dorcas • Jamaican Mestra
Long Bay, Portland parish, Jamaica

Nymphalidae: Biblidinae
Hamadryas februa • Gray Cracker
Gamboa, Colón, Panama

Nymphalidae: Biblidinae
Nica flavilla • Little Banner
Playa Los Angeles, Magdalena, Colombia

Nymphalidae: Biblidinae
Hamadryas laodamia • Starry Night Cracker ♀
Parque Nacional Pico Bonito, Atlántida, Honduras

Nymphalidae: Biblidinae
Pyrrhogyra cf. otolais • Common Redring
Shintuya Hot Springs, Madre de Dios, Peru

Nymphalidae: Biblidinae
Pyrrhogyra sp. • Redring
Boca Colorado, Madre de Dios, Peru

Nymphalidae: Charaxinae
Fountainea johnsoni • Johnson's Leafwing
Long Bay, Portland parish, Jamaica

Nymphalidae: Biblidinae
Temenis laothoe • Orange Banner
Shintuya Hot Springs, Madre de Dios, Peru

Nymphalidae: Charaxinae
Memphis philumena indigotica • Indigo Streaky Leafwing
Tuis, Cartago, Costa Rica

Nymphalidae: Charaxinae
Archaeoprepona demophon • One-spotted Prepona
Parque Nacional Pico Bonito, lodge, Atlántida, Honduras

Nymphalidae: Satyrinae
Calisto confusa or C. debarriera • Confused Calisto
Maniche, département Sud, Haiti

Nymphalidae: Satyrinae
Chloreuptychia agatha • Agatha Blue Ringlet
Montsinéry, Montsinéry-Tonnegrande, French Guiana

Nymphalidae: Satyrinae
Chloreuptychia cf. herseis • Godart's Blue Ringlet
Along the Rio Negro, Amazonas, Brazil

Nymphalidae: Satyrinae
Chloreuptychia sp. • Blue Ringlet
Montsinéry, Montsinéry-Tonnegrande, French Guiana

Nymphalidae: Satyrinae
Caeruleuptychia coelestis • Heavenly Blue Satyr
Lago Machuwasi, Madre de Dios., Peru

Nymphalidae: Satyrinae
Chloreuptychia chlorimene • Green Moon Ringlet
Along the Rewa River, Guyana

Nymphalidae: Satyrinae
Cissia penelope • Penelope's Ringlet
Berg en Dal, Brokopondo, Suriname

Nymphalidae: Satyrinae
Cissia pompilia • Pompilia Satyr
Selva Negra, Matagalpa, Nicaragua

Nymphalidae: Satyrinae
Cissia cf. terrestris • Terrestrial Satyr
Novo Airão, Amazonas, Brazil

Nymphalidae: Satyrinae
Cissia pseudoconfusa • Gold-stained Satyr
Cerro Azul, Panamá, Panama

Nymphalidae: Satyrinae
Haywardella edmondsii • Edmonds's Satyr
Puerto Madryn, Chubut, Argentina

Nymphalidae: Satyrinae
Cissia themis • Themis Satyr
Naranjo, Alajuela, Costa Rica

Nymphalidae: Satyrinae
Pierella astyoche • Astyoche Pierella
Reserva Particular do Cristalino, Mato Grosso, Brazil

Nymphalidae: Satyrinae
Pierella helvina • Tawny Pierella
Tuis, Cartago, Costa Rica

Nymphalidae: Satyrinae
Taygetis sosis • Sosis Satyr
Along the Rewa River, Guyana

Nymphalidae: Satyrinae
Pierella helvina • Tawny Pierella
Tuis, Cartago, Costa Rica

Nymphalidae: Satyrinae
Taygetis uzza • Uzza Satyr
Comarca Indígena Emberá-Wounaan, Panama

Nymphalidae: Satyrinae
Pierella hyalinus • Glassy Pierella
Along the Rupununi River, Guyana

Nymphalidae: Satyrinae
Satyrotaygetis satyrina • Wide-bordered Satyr
Selva Negra, Matagalpa, Nicaragua

Nymphalidae: Satyrinae
Cithaerias pireta • Pink Glasswing
Selva Negra, Matagalpa, Nicaragua

Nymphalidae: Satyrinae
Pareuptychia ocirrhoe • Two-banded Satyr
Santo Domingo de los Colorados, Pichincha, Ecuador

Nymphalidae: Satyrinae
Hermeuptychia hermes • Hermes Satyr
Tuis, Cartago, Costa Rica

Nymphalidae: Morphinae
Caligo uranus • Uranus Owl
Parque Nacional Pico Bonito, Atlántida, Honduras

Nymphalidae: Satyrinae
Euptychia cf. mollina • Mollina Satyr
Montsinéry, Montsinéry-Tonnegrande, French Guiana

Nymphalidae: Morphinae
Caligo telamonius • Yellow-fronted Owl
Tuis, Cartago, Costa Rica

Nymphalidae: Morphinae
Morpho helenor • Helena Morpho
Comarca Indígena Emberá-Wounaan, Panama

Nymphalidae: Ithomiinae
Greta annette • White-spotted Clearwing
Selva Negra, Matagalpa, Nicaragua

Nymphalidae: Morphinae
Morpho menelaus • Menaleus Morpho
Sentier de Bagne des Annamites, French Guiana

Nymphalidae: Ithomiinae
Heterosais giulia • Giulia Clearwing
Playa Los Angeles, Magdalena, Colombia

Nymphalidae: Morphinae
Morpho deidamia • Deidamia Morpho
Berg en Dal, Brokopondo, Suriname

Nymphalidae: Ithomiinae
Ithomia terra • Beautiful Glasswing
Nanegalito, Pichincha, Ecuador

Nymphalidae: Ithomiinae
Mechanitis polymnia • Orange-spotted Tiger Clearwing
Berg en Dal, Brokopondo, Suriname

Nyphalidae: Danainae
Danaus plexippus • Monarch
Santo Domingo de los Colorados, Pichincha, Ecuador

Nymphalidae: Ithomiinae
Oleria paula • Paula's Clearwing
Amatlán de los Reyes, Veracruz, Mexico

Nyphalidae: Danainae
Danaus gilippus • Queen
Crooked Tree Preserve, Belize district, Belize

Nymphalidae: Danainae
Danaus erippus • Southern Monarch
Zagarzazú, Colonia, Uruguay

Elachistidae: Stenomatinae
Stenoma byssina • Grass Miner Moth
Tuis, Cartago, Costa Rica

Euteliidae: Euteliinae
Paectes circularis • Circular Paectes
Montagne de Kaw, Roura, French Guiana

Hesperiidae: Eudaminae
Urbanus chalco • Great White Longtail
Sansón, Darién, Panama

Hesperiidae: Eudaminae
Epargyreus exadeus • Broken Silverdrop
Monteverde, Puntarenas, Costa Rica

Hesperiidae: Eudaminae
Urbanus dorantes • Dorantes Longtail Skipper
Paramaribo, Paramaribo district, Suriname

Hesperiidae: Eudaminae
Hyalothyrus neleus • Dimorphic Skipper
Comarca Indígena Emberá-Wounaan, Panama

Hesperiidae: Eudaminae
Urbanus simplicius • Plain Longtail Skipper
Novo Airão, Amazonas, Brazil

Hesperiidae: Eudaminae
Urbanus esmeraldus • Emerald Longtail Skipper
Playa Los Angeles, Magdalena, Colombia

Hesperiidae: Eudaminae
Phocides polybius • Guava Skipper
Crooked Tree Preserve, Belize district, Belize

Hesperiidae: Eudaminae
Urbanus viterboana • Bluish Longtail Skipper
Monteverde, Puntarenas, Costa Rica

Hesperiidae: Hesperiinae
Lerema accius • Clouded Skipper
Santa Elena, Puntarenas, Costa Rica

Hesperiidae: Eudaminae
Astraptes alardus • Frosted Flasher
La Suiza, Cartago, Costa Rica

Hesperiidae: Hesperiinae
Papias cf. *subcostulata* • Ruby-eye Skipper
Novo Airão, Amazonas, Brazil

Hesperiidae: Hesperiinae
Perichares philetes complex • Green-backed Ruby-eye
Nanegalito, Pichincha, Ecuador

Hesperiidae: Hesperiinae
≈ Talides sp. • Grass Skipper
Montsinéry, Montsinéry-Tonnegrande, French Guiana

Hesperiidae: Hesperiinae
Poanes zabulon • Zabulon Skipper
Monteverde, Puntarenas, Costa Rica

Hesperiidae: Hesperiinae
Vehilius stictomenes • Grass Skipper
Berg en Dal, Brokopondo, Suriname

Hesperiidae: Hesperiinae
Remella remus • Black-spotted Remella
Leona Vicario, Quintana Roo, Mexico

Hesperiidae: Hesperiinae
Callimormus radiola • Radiant Skipper
Tuis, Cartago, Costa Rica

Hesperiidae: Hesperiinae
Vettius marcus • Grass Skipper
Playa Los Angeles, Magdalena, Colombia

Hesperiidae: Hesperiinae
Anthoptes epictetus • Trailside Skipper
Novo Airão, Amazonas, Brazil

Hesperiidae: Hesperiinae
Hylephila phyleus • Fiery Skipper
Novo Airão, Amazonas, Brazil

Hesperiidae: Pyrginae
Antigonus erosus • Common Spurwing
Lago Bayano, Panamá province, Panama

Hesperiidae: Pyrginae
Cycloglypha thrasibulus • Widespread Bentwing
Boca Colorado, Manú Province, Peru

Hesperiidae: Pyrginae
Entheus matho • Matho Skipper
Gamboa, Colón, Panama

Hesperiidae: Pyrginae
Gorgythion begga • Variegated Skipper
Sansón, Darién, Panama

Hesperiidae: Pyrginae
Burnsius orcus • Orcus Checkered Skipper
Beside Lago Bayano, Panamá, Panama

Hesperiidae: Pyrginae
Heliopetes arsalte • Veined White-skipper
Along the Rupununi River, Guyana

Hesperiidae: Pyrginae
Pyrrhopyge phidias • Common Firetip
Limonal, Manú National Park, Madre de Dios, Peru

Hesperiidae: Pyrginae
Burnsius orcus • Orcus Checkered Skipper
Atalaya, Madre de Dios Department, Peru

Hesperiidae: Pyrginae
Anastrus meliboea • Frost-bordered Bluevent
Leona Vicario, Quintana Roo, Mexico

Hesperiidae: Pyrginae
Bolla cf. cupreiceps • Golden-headed Sootywing
Playa Los Angeles, Magdalena, Colombia

Hesperiidae: Eudaminae
Polythrix caunus • Longtail Skipper larva
Leona Vicario, Quintana Roo, Mexico

Limacodidae: Limacodinae
Acharia sp.• Slug Moth larva
Novo Airão, Amazonas, Brazil

Nymphalidae: Satyrinae
Caligo cf. telamonius • Giant Owl larva
Berg en Dal, Brokopondo, Suriname

Papilionidae: Papilioninae
Parides sp. • Cattleheart larva
Parque Nacional Pico Bonito, Atlántida, Honduras

Apatelodidae
Prothysana felderi • Shag Carpet Caterpillar
Tuis, Cartago, Costa Rica

Saturniidae: Hemileucinae
Automeris metzii • Metz's Eyed Silkmoth larva
Sansón, Darién, Panama

Sphingidae: Macroglossinae
Eumorpha nr. pandorus •Eumorpha Sphinx Moth larva
Tuis, Cartago, Costa Rica

Sphingidae: Macroglossinae
Xylophanes chiron • Chiron Sphinx larva
Along Rio Negro, Amazonas, Brazil

Nymphalidae: Satyrinae
Haetera piera • Amber Glasswing
Along the Rio Negro, Amazonas, Brazil

12

ORDER HYMENOPTERA

Sawflies, Wasps, Bees, and Ants

Hymenoptera is a remarkable order in many ways. It contains more social insects than any other order (eusocial is a more specific term). It contains more parasitoids than any other. Human fears of insects focus primarily on this sometimes-venomous order. But most importantly, its members are more responsible for pollination of the world's crops and other plants than any other insect group.

Order Hymenoptera is divided into two suborders. The smaller is Symphata, which contains 14 families of sawflies. The larger is Apocrita, made up of the wasps, bees, and ants. There are seven bee families, only one family of ants (Formicidae), and about 85 families of wasps. Note that with both bees and wasps some groups are eusocial while others are solitary. All ants are eusocial, while no sawflies are.

Anatomy. Hymenopterans can have the mouthparts directed forward or downward (i.e.prognathous or hypognathous, respectively). The basic form of the mouthparts in the order is mandibulate. In Apocrita the mandibles are often used for purposes other than chewing—in nest-building, for example. The compound eyes are typically large, and ocelli are present in some groups but not others. Antennae are long and have many antennomeres.

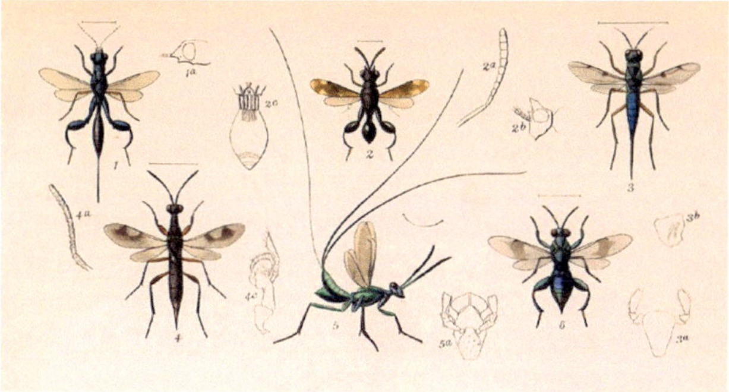

Figure 52 One of the many Hymenoptera plates from the massive publication *Biologia Centrali-Americana* (1879-1915).

The wings have reduced venation. The hindwings are joined to the forewings in flight by hooks called hamuli. Some Hymenoptera (worker ants, for example) are wingless.

Symphata have conventional divisions between the segments of the thorax and those of the abdomen. In Apocrita, on the other hand, the first segment of the abdomen is joined to the thorax, and what is called the thorax in other groups is termed the mesosoma. Also in Apocrita, segment two of the abdomen (and sometimes three) is greatly constricted into a "wasp waist" called the petiole. The petiole gives the abdomen increased flexibility, so that the ovipositor may be aimed in many different directions. In some groups the ovipositor evolved to inject venom, at first into hosts for parasitoidism, and later (in some groups) to be a stinger used in defense.

Food. Sawfly adults feed primarily on pollen and nectar, as do bees. Adult solitary wasps feed primarily on nectar. The various taxa of social wasps feed as adults on a variety of foodstuffs including nectar, fermenting fruit, insect prey, and scavenged dead arthropods.

Most ants are predators and/or scavengers. Leaf-cutter ants engage in agriculture, growing fungus on a substrate of leaf pieces. Other ants feed on mushrooms as a part of their diet. Some ants and some bees feed on the honeydew produced by leafhoppers, aphids, and other Auchenorrhyncha and Sternorrhyncha.

Reproduction. Hymenoptera have an unusual though not unique reproductive system called haplodiploidy. In this system the sex of the offspring depends on whether the egg was fertilized. Fertilized eggs produce females, while unfertilized ones produce males. A female can control whether or not she releases the sperm she has stored in order to fertilize the eggs and produce males. In many groups, such as ants, females greatly outnumber the males. In a small number of taxa males are unknown.

Development. The larvae of Symphata and of Apocrita are quite different from each other. Larvae in Symphata resemble caterpillars and have three thoracic legs as well as prolegs on most or all abdominal segments. The head capsule is well developed. In Apocrita the larvae are legless and have a reduced head capsule, though with fully developed mandibles.

In some Hymenoptera the larvae vary markedly by instar, a trait called heteromorphosis.

The larval food and feeding methods in Hymenoptera are also various. The larvae of Symphata are usually phytophagous (plant-feeding). Some Hymenoptera species induce galls wherein the larvae can feed and develop. Many Apocrita larvae develop as parasitoids of other insects, either internally, or attached to the outside of the host's integument. In some groups within Apocrita caregivers feed larvae with macerated prey, or with pollen and nectar.

Figure 53 *Pheidole biconstricta* ants eagerly feed on a mushroom, the bridal veil stinkhorn *(Phallus indusiatus).* Mato Grosso, Brazil.

Figure 54 A colorful sweat bee *(Pseudaugochlora sp.)* has fallen prey to a crab spider. El Salvador.

Figure 55 This small spider has managed to capture a much larger, venom-armed paper wasp. Suriname.

Dangers for Hymenoptera. Despite their strong mandibles and various chemical defenses including venom, Hymenoptera are preyed upon by birds, spiders, anteaters, skunks, mice, lizards, and other creatures. A few taxa of insects can develop only in the nests of Hymenoptera, where they plunder the food supply or kill and eat the larvae. Though a majority of species in the order are parasitoids, hymenopterans can also fall victim to parasitoids. Many tachinid flies, for instance, are parasitoids on sawfly larvae. Beehives can be attacked by varroa mites and tracheal mites that feed on bees' haemolymph and that can decimate a colony.

Hymenoptera are also vulnerable to bacterial and viral diseases. For example, beehives can be infested by a bacterium, *Paenibacillus larvae*, that infects larval bees.

Defenses. Their stings are the most important defenses of social wasps, bees, and many ant species including fire ants and bullet ants. Stings also protect velvet ants, which are actually solitary wasps in the family Mutillidae. Since stingers are modified ovipositors, only females can sting, and the many species that still use ovipositors for ovipositing cannot sting.

As in other orders, flying is a key defense of Hymenoptera. Mandibles in most groups are used for defense as well.

Some species, especially wasps, construct nests that provide protection from weather as well as from predators. Other species shelter underground. Groups such as hornets and army ants benefit from strength in numbers. A predator might be willing to risk receiving one sting in attacking a solitary wasp species, for example, but would be less likely to risk receiving a dozen stings by attacking a nest or colony of wasps, bees, or ants.

Few Hymenoptera make use of cryptic coloration. They are much more likely to sport aposematic coloration that warns predators of venom or other chemical defenses.

Relationship to humans. Sawflies, wasps, bees, and ants are economically important to humans, sometimes on the positive side of the ledger and sometimes on the negative side.

Humans view almost any hymenopterans living near their houses as pests and take steps to eliminate them. Many species of ant, for example, are seen as kitchen pests, while venomous creatures like fire ants and paper wasps are feared for their stings.

Being primarily phytophagous, sawfly larvae are considered enemies of farmers and gardeners. A number of these larvae feed on various grain crops, while others attack ornamental shrubs. On the other hand, the larvae of many wasp species are parasitoids on crop-eating caterpillars, and some are used in biological control.

While honey production is worth mentioning as a positive contribution to human society, much more important are bees' pollination services. No creature is more important a pollinator than the honey bee *(Apis mellifera),* but other bees such as bumble bees and leaf-cutting bees are also important in pollinating crops. Wasps, ants, and sawflies are important pollinators too. Overall, the order is the most important group of pollinators in the animal kingdom.

Neotropical Specialties. A large and varied fauna of paper wasps is widespread in Latin America. *Polistes* is one common genus found not only in Latin America but worldwide. Three other interesting genera are found only in the Neotropics. *Synoeca spp.* are commonly called "drumming wasps" because of the noise they make to warn predators. *Apoica* wasps are often pale or even whitish, their coloration presumably related to their nocturnal lifestyle. They boast large, prominent ocelli. *Polybia* is the most common genus of social wasps in the Neotropics. Species of *Polybia* can be hard to differentiate, and nest shape is often an aid in identification.

Figure 56 The structure shown here is an extrafloral nectary; it is on an inga tree. These nectaries provide ants with an easy way to feed on nectar. For their part, ants protect the tree from harmful insects such as gall-forming wasps. Brazil.

Among Neotropical species of bees, perhaps most noteworthy are the stingless bees that form tribe Meliponini. As the common name suggests, these bees do not sting, though they will bite in defense of the nest. The bites are not as painful to humans as bee stings. Aside from their stingless qualities, these bees are notable for making honey. For centuries native peoples have enjoyed the honey of domesticated stingless bees. Meliponine honey brings a high price at health food stores.

Figure 17 These stingless bees *(Trigona sp.)* are attentively watching over a group of adult and larval treehoppers *(Guayaquila pallescens)*. The treehoppers will provide the bees with honeydew. Honduras.

The ant fauna of the Neotropics is made up of many unusual groups. Trap-jaw ants (*Odontomachus spp.*) have mouthparts resembling a trap that can spring shut on prey with considerable force. A widespread species is the kelep (*Ectatomma tuberculatum*), a hunting ant found in many habitat types from Mexico to southern Brazil. Army ants *(Eciton spp.)* are remarkable for their foraging expeditions in which they attack numerous prey *en masse*. Fire Ants *(Solenopsis spp.)* are noted for the pain of their sting, although that of the bullet ant (*Paraponera clavata)* is much worse. Leafcutter ants *(Atta spp. and Acromyrmex spp.)* were mentioned above; they cut pieces of leaf and carry them back to the nest. The bits of leaf furnish the substrate for a fungus garden which feeds the colony. Of the ant genera mentioned, the ones endemic to Latin America are *Ectatomma, Eciton,* and *Paraponera*. *Solenopsis* ants are found worldwide. Trap-jaw and leaf-cutter ants are distributed from the far south of the United States, east to the Antilles, and south to Argentina.

Figure 58 This leaf-cutter ant (Attini) is cutting a leaf that it will soon carry back to the nest. There the piece will be used as a substrate for growing fungus. Paraguay.

Figure 58a Trap-jaw ants *(Odontomachus sp.)* "herding" treehopper nymphs *(Enchenopa sp.)* Suriname.

Figure 59 A gaggle of *Atta columbica* ants carry leaf pieces into their underground nest. Panama.

Figure 60 This gliding ant *(Cephalotes sp.)* is "shepherding" insects of two different groups: aphids and larval treehoppers. The ant, in turn, feeds on honeydew provided by the hemipterans. The ant's red abdomen indicates it is likely infested with nematodes.

Figure 61 Nests of paper wasps (Polistinae). **Top row**: *Polybia (Trichinothorax) sp.*, Cartago, Costa Rica. *Chartergellus sp.,* Colonia, Uruguay. *Polybia cf. bistriata*, Brokopondo, Suriname. **Bottom row**: *Parachartergus sp.,* Veracruz, Mexico. *Synoeca septentrionalis*, Veracruz, Mexico. Epipona sp., Madre de Dios, Peru.

Tenthredinidae: Blennocampinae
Waldheimia sp. • Sawfly
Rewa village, Upper Takutu-Upper Essequibo, Guyana

Tenthredinidae: Selandriinae
≈ Proselandria sp. • Sawfly
Tuis, Cartago, Costa Rica

Tenthredinidae: Blennocampinae
Waldheimia ochra or W. pallens • Sawfly
Berg en Dal, Brokopondo, Suriname

Tenthredinidae: Selandriinae
Selandriinae (poss. Stromboceridea sp.) • Sawfly
Tuis, Cartago, Costa Rica

Tenthredinidae: Heterarthrinae
Notofenusa sp. • Sawfly
Reserva Forestal Magallanes, Punta Arenas, Chile

Stephanidae: Stephaninae
Stephaninae • Crown Wasp
Tuis, Cartago, Costa Rica

Evaniidae: Evaniinae
Evaniella sp. • Ensign Wasp
Along Arroyo de las Víboras, Colonia, Uruguay

Diapriidae: Diapriinae
Spilomicrini • Diapriine Wasp
Boquete, Chiriquí, Panama

Trigonalidae: Trigonalinae
Taeniogonalos ornata • Trigonalid Wasp
El Hato, Sacatepéquez, Guatemala

Platygastridae: Scelioninae
Chromoteleia sp. • Scelionine Wasp
Selva Negra, Matagalpa, Nicaragua

Diapriidae: Diapriinae
Trichopria sp. • Diapriine Wasp
Boquete, Chiriquí, Panama

Platygastridae: Scelioninae
Scelioninae • Scelionine Wasp
Parque Nacional Tayrona, Magdalena, Colombia

Platygastridae: Scelioninae
Scelioninae • Scelionine Wasp
Tuis, Cartago, Costa Rica

Eulophidae
Eulophidae • Eulophid Wasp
Parque Nacional Tayrona, Magdalena, Colombia

Eucharitidae
Eucharitidae • Eucharitid Wasp
Parque Nacional Pico Bonito, lodge, Honduras

Pteromalidae: Diparinae
Diparinae • Diparine Wasp
Mountain Pine Ridge Forest Reserve, Belize

Eucharitidae: Oraseminae
≈ Orasema sp. • Eucharitid Wasp
Tuis, Cartago, Costa Rica

Eupelmidae: Neanastatinae
Neanastatinae • Neanastatine Wasp
Tuis, Cartago, Costa Rica

Eupelmidae: Neanastatinae
Metapelma sp. • Neanastatine Wasp
Tuis, Cartago, Costa Rica

Ichneumonidae: Cryptinae
Cryptini • Ichneumon Wasp
Montsinéry, Montsinéry-Tonnegrande, French Guiana

Ichneumonidae: Ichneumoninae
Joppa sp. • Ichneumon Wasp
Selva Negra, Matagalpa, Nicaragua

Ichneumonidae: Cryptinae
Cryptini • Ichneumon Wasp
Naranjo, Alajuela, Costa Rica

Ichneumonidae: Cryptinae
Cryptini • Ichneumon Wasp
Along the Rewa River, Guyana

Ichneumonidae: Cryptinae
Cryptini • Ichneumon Wasp
Tuis, Cartago, Costa Rica

Ichneumonidae: Ophioninae
Enicospilus nr. bozai • Ophionine Wasp
Montagne de Kaw, Roura, French Guiana

Ichneumonidae: Campopleginae
Campopleginae • Campoplegine Wasp
4 km north of Antigua, Guatemala

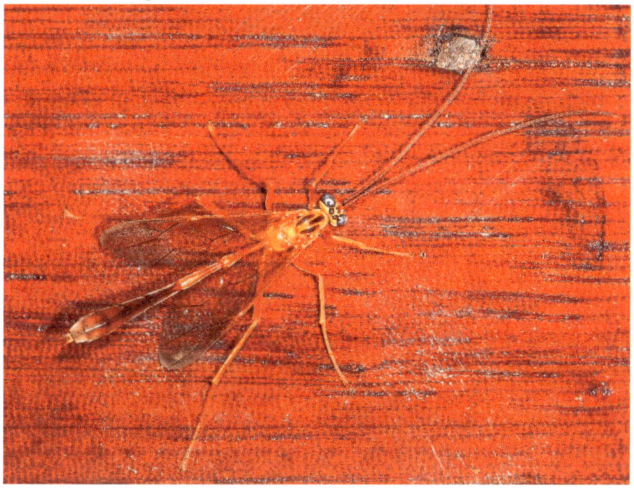

Ichneumonidae: Ophioninae
Enicospilus sp. • Ophionine Wasp
Berg en Dal, Brokopondo, Suriname

Ichneumonidae: Phygadeuontinae
Phygadeuontinae • Ichneumon Wasp
Montsinéry, Montsinéry-Tonnegrande, French Guiana

Ichneumonidae: Pimplinae
Xanthopimpla sp. • Pimpline Wasp
Selva Negra, Matagalpa, Nicaragua

Braconidae: Braconinae
Bracon sp. • Braconid Wasp
Selva Negra, Matagalpa, Nicaragua

Braconidae: Euphorinae
≈ Microctonus sp. • Euphorine Wasp
El Hato, Sacatepéquez, Guatemala

Braconidae: Rogadinae
≈ Bulborogas sp. • Braconid Wasp
Montagne de Kaw, Roura, French Guiana

Braconidae: Agathidinae
Zelomorpha sp. • Braconid Wasp
Montagne de Kaw, Roura, French Guiana

Chrysididae: Chrysidinae
Holophris cf. albolimbatus • Cuckoo Wasp
El Hato, Sacatepéquez, Guatemala

Braconidae: Microgastrinae
Microgastrinae • Microgastrine Wasp
Zitácuaro, Michoacán, Mexico

Rhopalosomatidae: Rhopalosomatinae
≈ Rhopalosoma sp. • Rhopalosomatid Wasp
Tuis, Cartago, Costa Rica

Thynnidae: Myzininae
Myzinum sp. • Myzinine Wasp ♂
Crooked Tree Preserve, Belize district, Belize

Mutillidae: Sphaeropthalminae
Hoplomutilla xanthocerata • Yellow-combed Velvet Ant
Altos del Maria, Coclé, Panama

Mutillidae: Sphaeropthalminae
Hoplomutilla sp. • Spherical-eyed Velvet Ant
Along the Rupununi River, Guyana

Mutillidae: Mutillinae
Timulla cf. mediata • Velvet Ant
Along the Suriname River, Brokopondo, Suriname

Mutillidae: Sphaeropthalminae
Hoplomutilla sp. • Spherical-eyed Velvet Ant
Pantiacolla, Madre de Dios, Peru

Mutillidae: Sphaeropthalminae
Pseudomethoca sp. • Velvet Ant
Tuis, Cartago, Costa Rica

Scoliidae: Campsomerinae

Pygodasis ephippium • Clear-winged Saddleback Wasp

Boquete, , Chiriquí, Panama

Pompilidae: Pepsinae

≈ Auplopus sp. • Pepsine Spider Wasp

Montsinéry, Montsinéry-Tonnegrande, French Guiana

Scoliidae: Campsomerinae

Campsomeris cf. trifasciata • Scoliid Wasp

Long Bay, Portland parish, Jamaica

Pompilidae: Pepsinae

Caliadurgus sp. • Pepsine Spider Wasp

Berg en Dal, Brokopondo, Suriname

Scoliidae: Scoliinae

Scoliinae • Scoliid Wasp

Santo Domingo de los Colorados, Pichincha, Ecuador

Pompilidae: Pepsinae

Pepsis sp. • Pepsine Spider Wasp

Boquete, Chiriquí, Panama

Pompilidae: Pepsinae
Pepsis sp. • Pepsine Spider Wasp
Suchitoto, Cuscatlán, El Salvador

Pompilidae: Pompilinae
Anoplius amethistynus • Blue-black Spider Wasp
Laguna de Oviedo, Pedernales, Dominican Republic

Pompilidae: Pepsinae
Priocnessus sp. • Spider Wasp
El Hato, Sacatepéquez, Guatemala

Pompilidae: Pompilinae
Poecilopompilus interruptus • Spider Wasp
Santo Domingo de los Colorados, Pichincha, Ecuador

Pompilidae: Pepsinae
Priocnemella cf. rufothorax • Spider Wasp
Along the Rewa River, Guyana

Pompilidae: Pompilinae
Poecilopompilus sp. • Spider Wasp
Montagne de Kaw, Roura, French Guiana

Vespidae: Eumeninae
≈ Omicron sp. • Potter Wasp
Along the Rupununi River, Guyana

Vespidae: Polistinae
Brachygastra cf. lecheguana • Honey Wasp
Colonia del Sacramento, Colonia, Uruguay

Vespidae: Eumeninae
Stenodynerus sp. • Potter Wasp
Parque Nacional Tayrona, Magdalena, Colombia

Vespidae: Polistinae
Clypearia cf. weyrauchi • Weyrauch's Paper Wasp
Tambo Blanquillo, Madre de Dios, Peru

Vespidae: Eumeninae
≈ Stenosigma sp. • Potter Wasp
Parque Nacional Tayrona, Magdalena, Colombia

Vespidae: Polistinae
Polistes crinitus • Paper Wasp
Long Bay, Portland parish, Jamaica

Vespidae: Polistinae
Polistes carnifex • Executioner Wasp
Parque Nacional Tayrona, Magdalena, Colombia

Vespidae: Polistinae
Polistes instabilis • Paper Wasp
Selva Negra, Selva Negra, Matagalpa, Nicaragua

Vespidae: Polistinae
Polistes cf. testaceicolor • Brick-colored Paper Wasp
Montagne de Kaw, Roura, French Guiana

Vespidae: Polistinae
Polistes myersi curassavicus • Curaçao Paper Wasp
Savaneta, Savaneta, Aruba

Vespidae: Polistinae
Polistes infuscatus • Darkened Paper Wasp
El Llano, Panamá province, Panama

Vespidae: Polistinae
Polistes pacificus • Paper Wasp
Selva Negra, Matagalpa, Nicaragua

Vespidae: Polistinae
Polistes versicolor • Versicolor Paper Wasp
Parque Nacional Tayrona, Magdalena, Colombia

Vespidae: Polistinae
Agelaia areata • Paper Wasp
El Hato, Sacatepéquez, Guatemala

Vespidae: Polistinae
Agelaia cf. myrmecophila • Paper Wasp
Novo Airão, Amazonas, Brazil

Vespidae: Polistinae
Agelaia panamaensis • Carnicera
El Hato, Sacatepéquez, Guatemala

Vespidae: Polistinae
Agelaia pallipes • Paper Wasp
Réserve Naturelle Trésor, Roura, French Guiana

Vespidae: Polistinae
Mischocyttarus sp. • Paper Wasp
Montsinéry, Montsinéry-Tonnegrande, French Guiana

Vespidae: Polistinae
Mischocyttarus cf. basimacula • Paper Wasp
El Hato, Sacatepéquez, Guatemala

Vespidae: Polistinae
Apoica pallens • Nocturnal Paper Wasp
Parque Nacional Pico Bonito, lodge, Atlántida, Honduras

Vespidae: Polistinae
Mischocyttarus sp. • Paper Wasp
Selva Negra, Matagalpa, Nicaragua

Vespidae: Polistinae
Apoica cf. strigata • Nocturnal Paper Wasp
Parque Nacional Tayrona, Magdalena, Colombia

Vespidae: Polistinae
Mischocyttarus cf. injucundus • Paper Wasp
Berg en Dal, Brokopondo, Suriname

Vespidae: Polistinae
Apoica thoracica • Nocturnal Paper Wasp
Berg en Dal, Brokopondo, Suriname

Vespidae: Polistinae
Parachartergus cf. apicalis • Paper Wasp
Selva Negra, Matagalpa, Nicaragua

Vespidae: Polistinae
Polybia (Myrapetra) sp. • Paper Wasp
Suchitoto, Cuscatlán department, El Salvador

Vespidae: Polistinae
Polybia sp. • Paper Wasp
El Hato, Sacatepéquez, Guatemala

Vespidae: Polistinae
Polybia (Trichinothorax) sp. • Paper Wasp
Tuis, Cartago, Costa Rica

Vespidae: Polistinae
Polybia cf. bistriata • Paper Wasp
Berg en Dal, Brokopondo, Suriname

Vespidae: Polistinae
Polybia occidentalis • Western Polybia
Montsinéry, Montsinéry-Tonnegrande, French Guiana

Vespidae: Polistinae
Polybia ruficeps • Red-footed Paper Wasp
Filadelfia, Boquerón, Paraguay

Vespidae: Polistinae
Protopolybia sedula • Paper Wasp
Montagne de Kaw, Roura, French Guiana

Vespidae: Polistinae
Polybia rejecta • Rejected Paper Wasp
Berg en Dal, Brokopondo, Suriname

Vespidae: Polistinae
Synoeca cf. surinama • Carton Wasps
Montagne de Kaw, Roura, French Guiana

Vespidae: Polistinae
Polybia scutellaris • Yellow-shielded Polybia
Punta Gorda, Nueva Palmira, Colonia, Uruguay

Vespidae: Polistinae
Synoeca virginea • Carton Wasp
Along the Rupununi River, Guyana

Ampulicidae: Ampulicinae
Ampulex sp. • Cockroach Wasp
Tuis, Cartago, Costa Rica

Sphecidae: Sceliphrinae
Sceliphron sp. • Mud Dauber
Lagunas de Aruza, Darién, Panama

Sphecidae: Sceliphrinae
Podiini • Thread-waisted Wasp
Nueva Palmira, Colonia, Uruguay

Sphecidae: Sphecinae
≈ Isodontia sp. • Grass-carrying Wasp
Long Bay, Portland parish, Jamaica

Sphecidae: Sceliphrinae
Sceliphron sp. • Mud Dauber
Novo Airão, Amazonas, Brazil

Sphecidae: Sphecinae
Sphex sp. • Digger Wasp
Puerto Madryn, Chubut, Argentina

Sphecidae: Ammophilinae
Ammophila sp. • Thread-waisted Wasp
Yumani, Isla de Sol, Lago Titicaca, Bolivia

Crabronidae: Bembicinae: Bembicin
≈ Bembix sp. • Sand Wasp
Paracas, Ica, Peru

Sphecidae: Ammophilinae
Eremnophila sp. • Thread-waisted Wasp
Leona Vicario, Quintana Roo, Mexico

Crabronidae: Bembicinae: Bembicini
Bicyrtes sp. • Sand Wasp
Puerto Madryn, Chubut, Argentina

Crabronidae: Bembicinae: Bembicini
Bembicini; prob. Stictia maculata • Sand Wasp
Montsinéry, Montsinéry-Tonnegrande, French Guiana

Crabronidae: Bembicinae: Bembicini
Trichostictia sp. • Sand Wasp
Paracas, Pisco Province, Peru

Crabronidae: Bembicinae: Bembicini
Bembicini; nr. Microbembex sp. • Sand Wasp
Paracas, Ica, Peru

Crabronidae: Crabroninae: Larrini
Subtribe Larrina • Square-headed Wasp
Berg en Dal, Brokopondo, Suriname

Crabronidae: Crabroninae: Larrini
Subtribe Gastrosericina • Square-headed Wasp
Berg en Dal, Brokopondo, Suriname

Crabronidae: Crabroninae
Trypoxylon sp. • Mud-dauber Wasp
Berg en Dal, Brokopondo, Suriname

Crabronidae: Crabroninae: Larrini
Subtribe Gastrosericina • Square-headed Wasp
Puerto Madryn, Chubut, Argentina

Crabronidae: Pemphredoninae
Pemphredoninae • Aphid Wasp
El Hato, Sacatepéquez, Guatemala

Colletidae: Colletinae
Colletes sp. • Cellophane Bee
Nueva Palmira, Colonia, Uruguay

Andrenidae: Panurginae
Pseudopanurgus sp. • Mining Bee
El Hato, Sacatepéquez, Guatemala

Colletidae: Diphaglossinae
Ptiloglossa sp. • Crepuscular Plasterer Bee
Crooked Tree Preserve, Belize district, Belize

Halictidae: Halictinae
Caenaugochlora cf. costaricensis • Sweat Bee
Suchitoto, Cuscatlán, El Salvador

Colletidae: Diphaglossinae
Ptiloglossa sp. • Crepuscular Plasterer Bee
Tuis, Cartago, Costa Rica

Halictidae: Halictinae
Lasioglossum (Dialictus) sp. • Sweat Bee
Paracas, Ica, Peru

Halictidae: Halictinae
Lasioglossum (Dialictus) sp. • Sweat Bee
Mountain Pine Ridge Forest Reserve, Belize

Halictidae: Halictinae
Augochloropsis sp. • Sweat Bee
Filadelfia, Boquerón, Paraguay

Halictidae: Halictinae
Lasioglossum (Dialictus) sp. • Sweat Bee
Parque Nacional Tayrona, Magdalena, Colombia

Halictidae: Halictinae
Augochloropsis sp. • Sweat Bee
Monteverde, Puntarenas, Costa Rica

Halictidae: Halictinae
Augochlora sp. • Sweat Bee
Colonia del Sacramento, Colonia, Uruguay

Megachilidae: Megachilinae
Hypanthidioides cf. colombiae • Leaf-cutter Bee
Parque Nacional Tayrona, Magdalena, Colombia

Megachilidae: Megachilinae
Coelioxys sp. • Cuckoo Leaf-cutter Bee
Nueva Palmira, Colonia, Uruguay

Apidae: Xylocopinae
Xylocopa sp. • Carpenter Bee
Colonia del Sacramento, Colonia, Uruguay

Apidae: Xylocopinae
Ceratina cf. chloris • Small Carpenter Bee
Berg en Dal, Brokopondo, Suriname

Apidae: Xylocopinae
Xylocopa mordax • Carpenter Bee
Les Cayes, département Sud, Haiti

Apidae: Xylocopinae
Ceratina sp. • Small Carpenter Bee
Filadelfia, Boquerón, Paraguay

Apidae: Xylocopinae
Xylocopa cf. mendozana • Mendoza Carpenter Bee
Filadelfia, Boquerón, Paraguay

Apidae: Apinae: Meliponini
Nannotrigona perilampoides • Nanotrigona nest entrance
Gamboa, Colón, Panama

Apidae: Apinae: Meliponini
Trigona sp. • Stingless Bee
Montsinéry, Montsinéry-Tonnegrande, French Guiana

Apidae: Apinae: Meliponini
Partamona cf. testacea • Stingless Bee nest entrance
Leaf-cutter ant nest below the bee nest entrance
Manú Road, Madre de Dios, Peru

Apidae: Apinae: Meliponini
Trigona sp. • Stingless Bees
Berg en Dal, Brokopondo, Suriname

Apidae: Apinae: Meliponini
Trigona sp. • Stingless Bee nest entrance
Gamboa, Colón, Panama

Apidae: Apinae: Meliponini
Trigona fulviventris • Stingless Bee

Suchitoto, Cuscatlán, El Salvador

Apidae: Apinae: Meliponini
Trigona muzoensis • Stingless Bee
Parque Nacional Pico Bonito, Honduras

Apidae: Apinae: Meliponini
Trigona pallens • Stingless Bees
Berg en Dal, Brokopondo, Suriname

Apidae: Apinae: Meliponini
Melipona fulva • Stingless Bee
Berg en Dal, Brokopondo, Suriname

Apidae: Apinae: Meliponini
Melipona (Michmelia) sp. • Stingless Bee
Valle de Antón, Coclé, Panama

Apidae: Apinae: Meliponini
Melipona sp. • Stingless Bee
Sentier de Bagne des Annamites, French Guiana

Apidae: Apinae: Meliponini
Partamona (Partamona) sp. • Stingless Bee
Selva Negra, Matagalpa, Nicaragua

Apidae: Apinae: Meliponini
Tetragona cf. clavipes • Stingless Bee
Montsinéry, Montsinéry-Tonnegrande, French Guiana

Apidae: Apinae: Centridini
Centris sp. • Oil Bee
Colonia del Sacramento, Colonia, Uruguay

Apidae: Apinae: Apini
Apis mellifera • Honey Bee
Savaneta, Savaneta, Aruba

Apidae: Apinae: Centridini
Centris mixta • Oil Bee
Paracas, Ica, Peru

Apidae: Apinae: Exomalopsini
Exomalopsis cf. bruesi • Long-horned Bee
Paracas, Ica, Peru

Apidae: Apinae: Euglossini
Eufriesea sp. • Orchid Bee
Novo Airão, Amazonas, Brazil

Apidae: Apinae: Euglossini
Euglossa sp. • Orchid Bee
Crooked Tree Preserve, Belize district, Belize

Apidae: Apinae: Euglossini
Euglossa sp. • Orchid Bee
Valle de Antón, Coclé, Panama

Apidae: Apinae: Euglossini
Euglossa sp. • Orchid Bee
Monteverde, Puntarenas, Costa Rica

Apidae: Apinae: Bombini
Bombus dahlbomii • Giant Bumblebee
Parque Nacional Torres del Paine, Chile

Apidae: Apinae: Bombini
Bombus ephippiatus • Saddled Bumble Bee
Amatlán de los Reyes, Veracruz, Mexico

Apidae: Apinae: Bombini
Bombus funebris • Mourning Bumble Bee
Yumani, Isla del Sol, Lago Titicaca, Bolivia

Apidae: Apinae: Bombini
Bombus terrestris • Buff-tailed Bumblebee
Torres del Paine National Park, Última Esperanza, Chile

Apidae: Apinae: Eucerini
Florilegus cf. purpurascens • Orchid Bee
Paracas, Ica, Peru

Apidae: Apinae: Euglossini
Eulaema marcii or E. pseudocingulata • Orchid Bee
Novo Airão, Amazonas, Brazil

Formicidae: Ponerinae
Neoponera villosa • Hairy Neoponera Ant
Berg en Dal, Brokopondo, Suriname

Apidae: Apinae: Euglossini
Eulaema polychroma • Orchid Bee
Naranjo, Alajuela, Costa Rica

Formicidae: Ponerinae
Odontomachus sp. • Trap-jaw Ant
Réserve Naturelle Trésor, Roura, French Guiana

Formicidae: Ponerinae
Odontomachus sp. • Trap-jaw Ant
Berg en Dal, Brokopondo, Suriname

Formicidae: Ponerinae
Paraponera clavata • Bullet Ant
Manú Road, Madre de Dios, Peru

Formicidae: Ponerinae
Pachycondyla cf. *theresiae* • Ponerine Ant
Mountain Pine Ridge Forest Reserve, Belize

Formicidae: Ponerinae
Ponerini • Ponerine Ant
Montagne de Kaw, Roura, French Guiana

Formicidae: Ponerinae
Pachycondyla harpax • Rapacious Panther Ant
Altos del Maria, Coclé, Panama

Formicidae: Ecitoninae
Eciton hamatum • Hooked Army Ant
Tuis, Cartago, Costa Rica

Formicidae: Ecitoninae
Neivamyrmex pilosus • Army Ant
Suchitoto, Cuscatlán, El Salvador

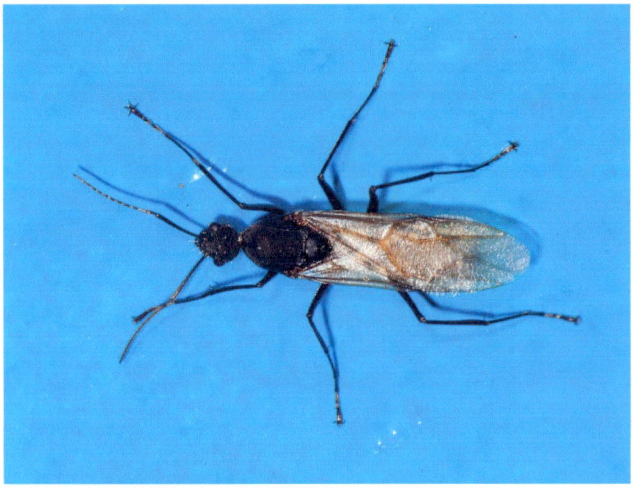

Formicidae: Formicinae
Camponotus sp. • Carpenter Ant ♂
Berg en Dal, Brokopondo, Suriname

Formicidae: Formicinae
Camponotus atriceps • Carpenter Ant
Novo Airão, Amazonas, Brazil

Formicidae: Formicinae
Camponotus cf. brettesi • Carpenter Ant
Gamboa, Colón, Panama

Formicidae: Formicinae
Camponotus sp. • Ground-dwelling Carpenter Ant
Isla del Sol, Lago Titicaca, Bolivia

Formicidae: Formicinae
Camponotus termitarius • Carpenter Ant
Yumani, Isla del Sol, Lago Titicaca, Bolivia

Formicidae: Formicinae
Gigantiops destructor • Giant-eyed Destroyer
Berg en Dal, Brokopondo, Suriname

Formicidae: Pseudomyrmecinae
Pseudomyrmex sp. • Twig Ant
Montsinéry, Montsinéry-Tonnegrande, French Guiana

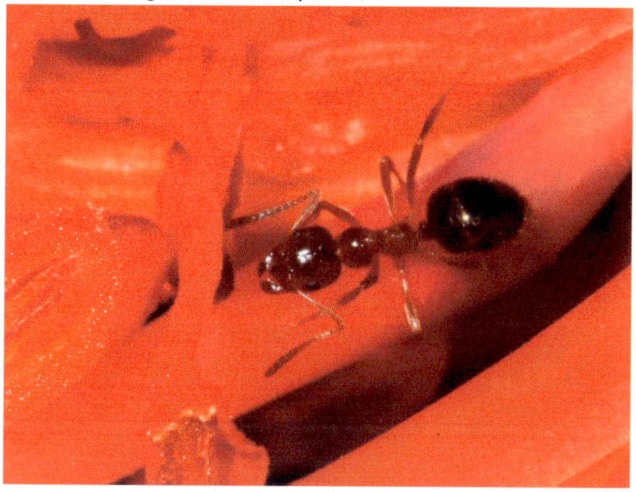

Formicidae: Formicinae
Lasiophanes piscinus • Formicine Ant
Última Esperanza, Chile

Formicidae: Pseudomyrmecinae
Pseudomyrmex sp. • Twig Ant
Altos del Maria, Coclé, Panama

Formicidae: Formicinae
Brachymyrmex patagonicus group • Brachymyrmex Ant
Filadelfia, Boquerón, Paraguay

Formicidae: Pseudomyrmecinae
Pseudomyrmex gracilis • Elongate Twig Ant
Along the Rupununi River, Guyana

Formicidae: Ectatomminae
Ectatomma brunneum • Ectatommine Ant
Novo Airão, Amazonas, Brazil

Formicidae: Myrmicinae
Crematogaster sp. • Acrobat Ant
Gamboa, Colón, Panama

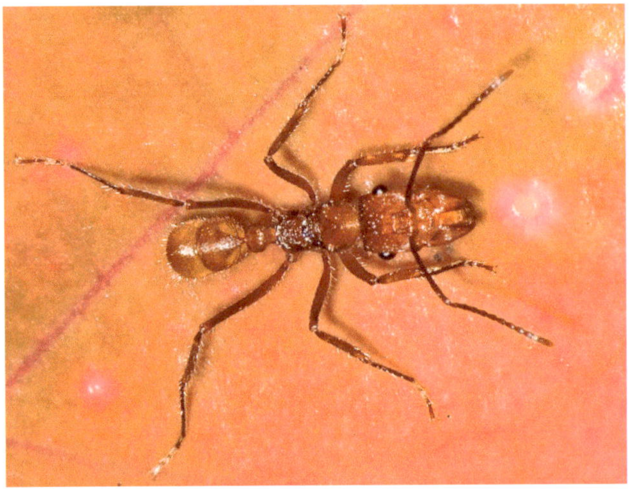

Formicidae: Ectatomminae
Ectatomma tuberculatum • Kelep
Colakreek, Para, Suriname

Formicidae: Myrmicinae
Crematogaster sp. • Acrobat Ant
Boquete, Chiriquí, Panama

Formicidae: Ectatomminae
Ectatomma sp. • Ectatommine Ant
Montsinéry, Montsinéry-Tonnegrande, French Guiana

Formicidae: Myrmicinae
Atta cephalotes • Leafcutter Ant
Suchitoto, Cuscatlán, El Salvador

Formicidae: Myrmicinae
Atta colombica • Colombian Leafcutter Ant
Gamboa, Colón, Panama

Formicidae: Myrmicinae
Acromyrmex sp. • Leaf-cutter Ants on a lemon
Colonia del Sacramento, Colonia, Uruguay

Formicidae: Myrmicinae
Atta sp. • Leafcutter Ants and paths they created
Parque Eco Arqueológico Los Naranjos, Honduras

Formicidae: Myrmicinae
Apterostigma sp. • Lesser Fungus-growing Ant ♂
Leona Vicario, Quintana Roo, Mexico

Formicidae: Myrmicinae
Acromyrmex sp. • Leaf-cutter Ant
Puerto Madryn, Chubut, Argentina

Formicidae: Myrmicinae
Trachymyrmex sp.• Lesser Fungus-Growing Ant ♀
Tortuguero National Park, Limón province, Costa Rica

Formicidae: Myrmicinae
Cephalotes atratus • Black-clothed Gliding Ant
Novo Airão, Amazonas, Brazil

Formicidae: Myrmicinae
Aphaenogaster araneoides • Spider-like Funnel Ants
Tuis, Cartago, Costa Rica

Formicidae: Myrmicinae
Pheidole biconstricta • Twice-constricted Myrmicine Ant
Along the Rio Negro, Amazonas, Brazil

Formicidae: Dolichoderinae
Dolichoderus attelaboides • Odorous Ant
Montsinéry, Montsinéry-Tonnegrande

Formicidae: Dolichoderinae
Dolichoderus cf. bispinosus • Two-spined Odorous Ant
Along the Rio Negro, Amazonas, Brazil

Formicidae: Dolichoderinae
Dolichoderus laminatus • Plated Dolichoderine Ant
Parque Nacional Tayrona, Magdalena, Colombia

13

OTHER INSECT ORDERS

Bristletails, Mayflies, Stoneflies, Earwigs, Webspinners, Bark Lice, Thrips

The last chapter of *Insects of Latin America* is a hodgepodge. It briefly examines some smaller orders including everything from the primitive bristletails to the more recently evolved thrips. The seven orders treated by this chapter do not have a lot in common, but none go through complete metamorphosis. They also have in common that compared with beetles or butterflies, they have far fewer experts devoted to them, and their life histories are often unknown.

Jumping Bristletails. Order Archaeognatha. This primitive, wingless group is unusual for continuing to grow and molt throughout its life, while other insects stop growing and molting on reaching the adult stage. In jumping bristletails the compound eyes are large and touch each other, at least in dorsal view. Three simple eyes (ocelli) are also present. The antennae are long and many-segmented, and the simple mandibles are partly withdrawn into the head. The thorax has a humped appearance in lateral view. Bristletails bear three terminal appendages: the shorter paired cerci, and a longer appendix dorsalis.

Males deposit droplets of sperm on or near the ground and the female then walks over the area and takes the sperm into her reproductive tract. Most bristletails bear slippery scales that help them escape predators. They can also run quickly. Flexing their humped thorax, they can jump impressive distances. Jumping bristletails live primarily in leaf litter and under bark. They are most active at night, feeding on mosses, algae, lichens, fungi, and the decaying leaves of flowering plants.

Mayflies. Order Ephemeroptera. Mayflies are more diverse in the Neotropics than in any other biogeographic region, with about 600 described species there. The compound eyes are large (especially in males) and three ocelli are present. The forewings are large and triangular, while the hindwings are smaller or even absent. When at rest the wings are held above the body, their tips pointed up. Males have long forelegs used to grab and grip the female in the mating flight. Both nymphs and adults have a pair of cerci at the end of the abdomen, and usually a centered terminal filament between the cerci. Adults have only vestigial mouthparts and do not feed, while the nymphs have fully developed mandibles. Anatomically, mayflies have changed little in the last 248 million years. Amazingly, the aquatic nymphs go through up to 45 instars. Another unusual aspect of nymphal development is that there is a subimago stage (pre-adult) which has milky wings; within a half-day or a day the subimago molts to the adult stage (with clear wings). Some adult mayflies live as little as an hour, their brief adult lives consisting of a mad scramble to mate. Except in the temperate areas such as southern Argentina and Chile, Neotropical mayflies emerge all year on a staggered basis, rather than in one great hatch as often happens in the Holarctic.

Stoneflies. Order Plectoptera. Stoneflies first arose many millions of years ago in the southern hemisphere, and today one of the two suborders (Antarctoperlaria) is found only in South America and Australasia. The Neotropics are home to about 500 described species of stonefly.

Stoneflies bear filiform antennae and mandibulate mouthparts. The compound eyes seem to bulge, and two or three ocelli are present. The hindwings are broader than the forewings, but both pairs are membranous, and the folded wings are held flat over the back when the insect is perched. Short-winged and wingless species do exist, however.

Stoneflies mate during daylight hours; the smaller male mounts the female. Eggs are laid in the water. Stonefly nymphs are aquatic and go through 10 to 24 instars. Nymphs feed on detritus, algae, various small water plants, and particularly in later instars they may be predacious or omnivorous. Adults feed on various kinds of plant matter and detritus; in some taxa adults do not feed.

Earwigs. Order Dermaptera. Earwigs are most noted for bearing "pincers" (enlarged cerci) at the apex of the abdomen. These are not strong enough to hurt humans. Their compound eyes vary in size among the taxa, and some species lack them. All earwigs lack ocelli. Many species also lack wings. Winged species bear short, leathery forewings that protect the larger membranous hindwings. Earwigs hold their wings flat over their backs, but both pairs are too short to completely hide the abdomen.

Earwigs mate tail-to-tail, and females lay eggs in the soil, leaf litter, or rotten wood. Females stay with the eggs, licking them occasionally to discourage the growth of fungus. The females may stay with the nymphs for the first two or three instars, even feeding them. After passing through four or five instars the earwigs reach the adult stage. Adult earwigs feed nocturnally on detritus and various kinds of plant materials, and they also prey on other arthropods.

Webspinners. Order Embioptera. Webspinners comprise a small order with less than 500 described species worldwide. In this order the compound eyes are kidney-shaped and ocelli are lacking. Females are wingless, while the males of most species have two pairs of narrow wings. Legs are short, and the hind femora noticeably thickened. These insects have a silk-producing gland on the enlarged first tarsal segment of the forelegs.

Webspinners produce silk-covered walkways leading out from a central retreat. These silken tunnels provide protection from predators and in drier areas help prevent desiccation. A typical colony consists of several adult females and their offspring. A male flies to the colony and locates a mate; he dies soon after mating. Nymphs go through four instars. Both nymphs and adult females have a diet of small bits of plant material and detritus. Adult males do not feed.

Bark Lice. Order Psocoptera. Bark lice and other non-parasitic lice number about 6000 named species worldwide. In these insects the compound eyes are large; ocelli are lacking in wingless species but three are present in winged ones. The mandibles are asymmetric. If present, the wings are membranous, with a reduced number of veins. Wings are held roof-like over the body.

Figure 61 These early instar insects are recognizable as bark lice (Psocoptera) by their long and slender antennae, swollen "forehead," and soft bodies. Bark lice have two large compound eyes and three ocelli. On the leaf of a *Solanum sp.*, Coclé, Panama.

Some species have elaborate courtship rituals. Males produce a spermatophore which the females take up. Nymphs pass through four instars in most bark lice. A few species always reproduce parthenogenetically, while others do so only in some circumstances. Food for adults and nymphs consists of tiny threads of fungus, fungal spores, algae, lichens, decaying deciduous leaves, and the like. Some bark lice live solitary lives, but some taxa are gregarious. In the latter case, the colony may be covered with a sheet of protective silk produced by the salivary glands.

Thrips. Order Thysanoptera. Thrips are very small, slender insects found on the leaves and flowers of plants and on fungi. Some species are wingless and others short-winged, while fully winged species have two pairs of slender, fringed wings. In suborder Terebrantia the wings are parallel when the insect is at rest, while in suborder Tubulifera they cross. The right mandible is vestigial and the left functions as piercing mouthpart. Thrips puncture their food, inject saliva, then suck up the liquified food.

Entomologists classify the reproduction of thrips as intermediate between hemimetabolous and holometabolous. These insects have two larval stages and two or three pupal stages. In common with the only distantly related Hymenoptera, females develop from fertilized eggs and males from unfertilized eggs. This means males have half the number of chromosomes as females have. Food for thrips include spores and filaments of fungi, as well as leaves and pollen. Some species are predatory. A few groups are garden pests, either inducing galls or carrying plant viruses. On the other hand, some thrips are valued as pollinators.

Machilidae
Machilidae ● Jumping Bristletail
Along the Rio Negro, Amazonas, Brazil

Baetidae: Baetinae
Baetinae • Small Mayfly nymph
Parque Nacional del Manú, Madre de Dios, Peru

Spongiphoridae: Spongiphorinae
Spongophora buprestoides • Spongiphorid Earwig
Tuis, Cartago, Costa Rica

Leptophlebiidae: Atalophlebiinae
Atopophlebia sp. • Prong-gilled Mayfly
Valle de Antón, Coclé, Panama

Spongiphoridae
Spongiphoridae • Spongiphorid Earwig
Monteverde, Puntarenas, Costa Rica

Perlidae: Acroneuriinae
Anacroneuria sp. • Common Stonefly
Mountain Pine Ridge Forest Reserve, Belize

Spongiphoridae: Spongiphorinae
Spongiphorinae • Spongiphorine Earwig
Tuis, Cartago, Costa Rica

Forficulidae: Ancistrogastrinae
Ancistrogaster sp. • Earwig
Nanegalito, Tandayapa Valley, Ecuador

Psocidae: Thyrsophorinae
Poecilopsocus cf. calocoroides • Bark Louse
Along the Rio Negro, Amazonas, Brazil

Oligotomidae
Oligotomidae • Webspinner (note swollen foretarsus)
Presqu'île de la Caravelle, Martinique

Psocidae: Psocinae
Cerastipsocus sp. • Bark Louse
Reserva Particular do Cristalino, Mato Grosso, Brazil

Epipsocidae
Goja sp. • Bark Louse
Boquete, Chiriquí province, Panama

Tubulifera
Tubulifera • Tubular Thrips
Selva Negra, Matagalpa, Nicaragua

BOOKS FOR FURTHER READING

Abadie E, Coelho Grossi P, Wagner P. 2008. Field guide to the Dynastidae family of the south of South America. Buenos Aires: n.p.

Bates HW. 1989. The naturalist on the river amazons: A record of adventures, habits of animals, sketches of Brazilian and Indian life, and aspects of nature under the equator, during eleven years of travel [1863]. New York: Penguin Books.

Bell WJ, Roth LM, Nalepa CA. 2007. Cockroaches: Ecology, behavior, and natural history. Baltimore: Johns Hopkins University Press.

Belt T. 1874. The naturalist in Nicaragua: A narrative of a residence at the gold mines of Chontales; journeys in the savannahs and forests, with observations on animals and plants in reference to the theory of evolution of living forms. London: J. Murray.

Blackwelder R. 1944-1957. Checklist of the coleopterous insects of Mexico, Central America, the West Indies, and South America. 6 vols. Washington: Smithsonian Institution.

Borror DJ. 1960. Dictionary of word roots and combining forms compiled from the Greek, Latin, and other languages, with special reference to biological terms and scientific names. Palo Alto, Calif.: N-P Publications.

Brown V, Borkent A, Cumming JM, Wood DM, Woodley NE, Zumbado M. 2010. Manual of Central American Diptera. 2 vols. . Ottawa: NRC Research Press.

Cadena-Castañeda O, Cardona-Granda JM. 2015. Introducción a los saltamontes de Colombia. 2 vols. Colombia: n.p.

Caranpaima JC. 2016. Grasshoppers of northwest South America: A photo guide. 2 vols. San Francisco: Blurb.

Carbonell CS. 2006. The Andes and Amazon: A naturalist's journal of a trip to the mountains and river in the year 1948. Montevideo: La Banda Oriental.

Castner J. 2000. Amazon insects: A photo guide. Gainesville, Florida: Feline Press.

Chacón I, Montero J. 2007. Mariposas de Costa Rica (orden Lepidoptera) = Butterflies and moths of Costa Rica (order Lepidoptera). Santo Domingo de Heredia, Costa Rica: INBio.

Conle O, Hennemann F, Gutiérrez Y. 2011. The stick insects of Colombia: A catalogue and bibliography with the descriptions of four new genera and 74 new species. Books on Demand GmbH.

Darwin C. 2001. The voyage of the Beagle: Journal of researches into the natural history and geology of the countries visited during the voyage of H.M.S. Beagle round the world [first published 1838, 1845]. New York: Modern Library.

de Almeida AC, Freitas AVL. 2012. Lepidoptera: Borboletas e mariposas do Brasil. São Paulo: Exclusiva Publicações Ltda.

Dett A. 2017. Moths of Costa Rica's rainforest. Benteli Verlags.

Erwin T. 2007-2011. A treatise on the western hemisphere Caraboidea (Coleoptera), their classification, distributions, and ways of life. 3 vols. Sofia: Pensoft. Second volume co-authored by DL Pearson.

Esquivel C. 2006. Libélulas de Mesoamérica y el Caribe = Dragonflies and damselflies of Middle America and the Caribbean. Santo Domingo de Heredia, Costa Rica: INBio.

Faust L. 2017. Fireflies, glow-worms, and lightning bugs. Athens, GA: University of Georgia Press.

Forsyth A. 2008. Nature of the rainforest: Costa Rica and beyond. Ithaca, NY: Comstock Publications.

Garrison RW, von Ellenrieder N, Louton JA. 2010. Damselfly genera of the new world: An illustrated and annotated key to the Zygoptera. Baltimore, Md.: Johns Hopkins University Press.

Garrison RW, von Ellenrieder N, Louton JA. 2006. Dragonfly genera of the new world: An illustrated and annotated key to the Anisoptera. Baltimore, Md.: Johns Hopkins University Press.

Garwood K, Lehman R. 2013. Butterflies of Central America. 3 vols. McAllen, Texas: RiCalé Publishing.

Garwood K, Lehman R. 2009. Butterflies of southern Amazonia. 2nd ed. McAllen, Texas: RiCalé Publishing.

Glassberg J. 2017. A swift guide to butterflies of Mexico and Central America. 2nd ed. Princeton, NJ: Princeton University Press.

Godoy C, Garnier X, Nishida K. 2006. Membrácidos de la América tropical = Treehoppers of tropical America. Santo Domingo de Heredia, Costa Rica: INBio.

Gowin E, Williams TT. 2017. Mariposas nocturnas: Moths of Central and South America : A study in beauty and diversity. Princeton: Princeton University Press.

Gullan PJ, Cranston PS. 2014. The insects: An outline of entomology. Fifth ed. Chichester, UK ; Hoboken, NJ: John Wiley & Sons, Inc.

Hanson P, Gauld I. 2006. Hymenoptera de la región Neotropical. Gainesville, Florida: American Entomological Institute.

Hanson PE, Nishida K. 2016. Insects and other arthropods of tropical America. Ithaca: Comstock Publishing.

Heckman CW. 2017. Encyclopedia of South American aquatic insects. Berlin: Springer.

Henderson CL. 2010. Butterflies, moths, and other invertebrates of Costa Rica: A field guide. Austin: University of Texas Press.

Hequet V. 1996. Longicornes de Guyane. Cayenne: Silvolab : ORSTOM.

Hogue CL. 1993. Latin American insects and entomology. Berkeley: University of California Press.

Hőlldobler B, Wilson EO. 2011. The leafcutter ants: Civilization by instinct. New York: Norton.

Jameson ML, Ratcliffe BC. 2011. The Neotropical scarab beetle tribe Anatistini. Lincoln, Nebraska: University of Nebraska State Museum.

Janzen D. 1983. Costa Rican natural history. Chicago: University of Chicago Press.

Kricher J. 1997. A Neotropical companion: An introduction to the animals, plants, and ecosystems of the new world tropics. 2nd ed. Princeton: Princeton University Press.

Lees DC. 2019. Moths: A complete guide to biology and behavior. Washington, DC: Smithsonian Books.

Marshall S. 2018. Beetles: The natural history and diversity of Coleoptera. Richmond Hill, ON: Firefly.

Marshall S. 2012. Flies: The natural history and diversity of Diptera. Richmond Hill, ON: Firefly.

Marshall S. 2008. 500 insects: A visual reference. Richmond Hill, Ontario: Firefly.

Martin A, Soares A, Bizarro J. 2011. Guia dos Sphingidae da Serra dos Órgãos, sudeste do Brasil = A guide to the hawkmoths of the Serra dos Orgaos, south-eastern Brazil. Oxford, UK: REGUA Publications; Information Press.

Naskrecki P. 2000. Katydids of Costa Rica, systematics and bioacoustics of the cone-head katydids. Philadelphia: Orthopterists' Society.

Naskrecki P, Wilson EO. 2017. Hidden kingdom: The insect life of Costa Rica. Ithaca: Comstock Publishing.

O'Toole C. 2002. Firefly encyclopedia of insects and spiders. Toronto: Firefly Books.

Otte D, Perez-Gelabert D. 2009. Caribbean crickets. N.p.: Orthopterists Society.

Paulson D, Haber W. 2021 [forthcoming]. Dragonflies and damselflies of Costa Rica: A field guide. Ithaca, NY: Comstock Publications.

Paulson D. 2019. Dragonflies & damselflies: A natural history. Princeton, NJ: Princeton University Press.

Peña G. LE, Ugarte P. AJ. 1996. Las mariposas de Chile = The butterflies of Chile. 1st ed. Santiago de Chile: Editorial Universitaria.

Pérez-Asso A, Genaro J, Bastardo R, López Sánchez A. 2017. Mariposas de la Española, butterflies of Hispaniola. N.p.: Editorial Cocuyo.

Purser B. 2003. Jungle bugs: Masters of camouflage and mimicry. Toronto: Firefly.

Panizzi A, Grazia J. 2015. True bugs (Heteroptera) of the Neotropics. Berlin: Springer.

Rowell CHF. 2013. The grasshoppers (Caelifera) of Costa Rica and Panama. N.p.: Orthopterists Society.

Tobar E. 2000. Coleopteros de Chile / Chilean beetles. Santiago: Fototeknika.

Volkmann L, Núñez Bustos E. 2013. Mariposas Serranas de Argentina central. Córdoba, Argentina: Equipo Gráfico.

Zumbado M. 2006. Dípteros de Costa Rica y la América tropical. Santo Domingo de Heredia, Costa Rica: INBio.

Index to Photographs

Abana, 128

Abracris, 29

Acanalonia, 114

Acanaloniidae, 114

Acanthagrion, 8, 20

Acanthonycha, 196

Acanthopidae, 52–53

Acharia, 314

Achilidae, 108–109

Aclodes, 37

Aconophora, 132

Acontista, 51, 53

Acraga, 249

Acrididae, 28–32

Acrocinus, 147

Acromyrmex, 352-353

Actinote, 293

Adeloneivaia, 256

Adhemarius, 258

Aegithus, 174

Aegorhinus, 204

Aellopos, 259

Aeolus, 165

Aeshnidae, 11

Aetalionidae, 136

Agalliota, 117

Agaraea, 272

Agelaia, 334

Agra, 154

Agraulis, 293

Agriacris, 33

Agrilaxia, 164

Agrilus, 164

Agriocoris, 81

Agroiconota, 188

Agrosoma, 124

Agrypnus, 165

Alagoasa, 197

Alchisme, 135

Allochroma, 196

Allominettia, 228

Alydidae, 89–91

Amapala, 111

Amarynthis, 291

Amastus, 279

Amaxia, 278-279

Amblyonychus, 218

Amblypsilopus, 221

Amblyscarta, 125

Amblytropidia, 31

Amitermes, 65

Ammophila, 338

Ammophilinae, 338-339

Ampulex, 338

Ampulicidae, 338

Anablysis, 29

Anacroneuria, 358

Anartia, 298

Anasa, 95

Anastrepha, 227

Anastrus, 313

Anatya, 11–12

Ancistrogastrinae, 360

Ancognatha, 162

Andrenidae, 341

Anisagrion, 20

Anisoscelis, 92

Anoba, 270

Anomala, 160

Anoplius, 331

Anteos, 285

Anteros, 292

Anthanassa, 296

Anthericonia, 42

Anthomyia, 231

Anthomyiidae, 231

Anthoptes, 312

Antianthe, 133

Antiblemma, 271

Anticarsia, 271

Anticla, 255

Antigonus, 312

Antiteuchus, 99

Antlions, 141, 144

Ants, 318-321, 348–354

Apatelodes, 254-255

Apatelodidae, 254–255

Apateticus, 102

Aphaenogaster, 353

Aphididae, 137
Aphids, 137
Aphleboderrhis, 84
Aphodiinae, 156
Apidae, 342–348
Apidaurus, 90
Apinae, 343–348
Apiomerinae, 80-81
Apiomerus, 80
Apis, 345
Apogonalia, 125
Apoica, 323-354, 335
Apotinocerus, 217
Apterostigma, 353
Aracynthus, 110
Aradidae, 84–85
Arawacus, 287
Archaeoprepona, 302
Archilestes, 9, 23
Arctiinae, 244, 272-281
Areva, 273
Argia, 18–20
Argyria, 254
Aricoris, 292
Ascalaphidae, 143, 146
Ascia, 287
Asilidae, 217–219, 234
Aslamidium, 188
Asphaera, 197
Aspidolea, 162
Aspisoma, 170
Astraptes, 310
Asturodes, 251
Astylus, 172
Ataenius, 156
Atalophlebiinae, 359
Ateuchus, 161
Atomosia, 219
Atoniomyia, 219
Atractocerus, 172
Atractosomus, 167
Atta, 321, 352
Atteva, 246
Attevidae, 246
Auchenorrhyncha, 4, 70,103-137
Augochlora, 341
Augochloropsis, 342
Auplopus, 330
Auratonota, 247
Automeris, 257, 315
Azeta, 271
Bactrodes, 83
Baetidae, 359
Baleja, 120
Baliopteridion, 228
Banasa, 98
Bandaromimus, 130
Baridinae, 206
Barybas, 159
Battus, 284
Bees, 340-348
Beetles, 147-206
Beirneola, 120
Belta, 200
Bembicinae, 339
Bembix, 339
Beraba, 183
Beridinae
Berosus, 154
Bertholdia, 274
Berytidae, 85
Bibio, 211
Bibionidae, 211
Bicellonycha, 171
Biolleyana, 111–112
Blaberidae, 67–68
Blaberus, 68
Bladina, 112
Blattidae, 66
Blondeliini, 232
Bocydium, 135
Bolbites, 162
Bolla, 314
Bombus, 347
Bombycidae, 255
Bombyliidae, 216–217
Bostra, 47
Bostrichidae, 172
Bothrideridae, 176
Bothriocera, 105
Brachygastra, 332
Brachymyrmex, 350
Brachystethus, 100

Bracon, 327
Braconidae, 327–328
Brasiella, 151
Brechmorhoga, 10, 13
Brenthia, 246
Brentidae, 201
Brentus, 201
Bristletails, jumping, 355, 357
Bryolymnia, 283
Bugs, true, 72, 74-103
Bulborogas, 328
Buprestidae, 164
Burnsius, 313
Burtinus, 90
Butterflies, 284-316
Byllis, 108
Cacocharis, 247
Caddisflies, 237-240
Caenaugochlora, 341
Caeruleuptychia, 303
Calephelis, 292-293
Caliadurgus, 330
Caligo, 306, 314
Calisto, 302
Calledema, 268
Callimormus, 311
Callionima, 259
Calliphoridae, 232
Callipogon, 182
Callopistria, 283
Calocladon, 167
Calocorisca, 76
Calondas, 77
Calonotos, 276
Calophaena, 153
Calopteron, 167
Calopterygidae, 17–18
Calospila, 291-292
Calycopis, 288-289
Calyptoproctus, 111
Camponotus, 349–350
Campopleginae, 327
Campsomeris, 330
Campylenchia, 132
Cannaphila, 13
Canopidae, 103
Canopus, 103
Cantharidae, 171–172
Canthidium, 157-158
Canthon, 157
Capraita, 198
Carabidae, 151–154
Cardiacephala, 224
Cardioscarta, 125
Carineta, 136
Castniidae, 248
Castniomera, 248
Castolus, 79-80
Cataulax, 99
Caterpillars, 314-315
Catorhintha, 91
Catrimania, 114
Caviria, 283
Cedusa, 105
Centris, 346
Cephalotes, 321, 353
Ceraeochrysa, 141
Ceraia, 37
Cerambycidae, 147–148, 182-186
Ceraspis, 156
Cerastipsocus, 360
Ceratina, 342
Ceratopogonidae, 213
Ceratopus, 205
Cercopidae, 114–117
Ceroctena, 283
Cerotoma, 192
Chaetominettia, 228
Chalcolepidius, 166
Chalcophana, 199
Chalepispa, 187
Chalepus, 187
Charaxinae
Charidotella, 191
Charidotis, 191
Chariesterus, 91
Chartergellus, 322
Chauliognathus, 171
Chelymorpha, 191
Chinaia, 131
Chioneinae, 211
Chironomidae, 213
Chloreuptychia, 303

Chlorida, 183
Chlorocoris, 99
Chlorogonalia, 125
Chloropteryx, 264
Chlorotabanus, 214
Chlosyne, 296-297
Choeradodis, 49, 59-60
Cholomyia, 233
Cholus, 204
Chordonota, 216
Choreutidae, 246
Chorisoneura, 66-67
Chromoteleia, 324
Chrysididae, 328
Chrysobothris, 164
Chrysomelidae, 186–200
Chrysoperla, 142
Chrysopidae, 141–142
Chrysopodes, 142
Cicadas, 73-74, 136-137
Cicadellidae, 117-131
Cicadidae 73-74, 136-137
Cicindela, 151
Cissia, 303–304
Cithaerias, 306
Cixiidae, 103–105
Cixius, 103
Cladonota, 134
Clastoptera, 117
Clastopteridae, 117
Clemensia, 274
Climaciella, 144
Clogmia, 212
Clypearia, 332
Cnemomis, 93
Coccinellidae, 177–178
Cockroaches, 63-69
Coelidiinae, 118–119
Coelioxys, 342
Coenagrionidae, 18–22
Coiffaitarctia, 280
Colaspis, 200
Coleomegilla, 177
Coleoptera, 147-206
Collabismus, 202
Collaria, 76
Colletes, 340
Colletidae, 340
Colobathristidae, 86
Colobura, 299
Colpolopha, 33
Compsacris, 32
Compsus, 201
Conchylodes, 251
Condylostylus, 221
Conocephalinae, 39-40
Conocephalus, 39
Copelatus, 154
Copestylum, 222
Copicerus, 107
Copidocephala, 109–110
Copiphora, 40
Cordyligaster, 233
Coreidae, 72, 91–96
Corematura, 244
Corimelaenidae, 103
Cornops, 28
Coronigoniella, 125-126
Correbia, 273
Corydalidae, 143, 145
Corydalus, 143, 145
Corynomalus, 178
Corythucha, 84
Cosmosoma, 277
Cossidae, 248
Crabro, 340
Crabronidae, 339–340
Crambidae, 242, 251–254
Cratomorphus, 170
Cratoplastis, 272
Cratosomus, 204
Crematogaster, 352
Crickets, 35-37
Crypticerya, 138
Cryptinae, 326
Ctenostoma, 152
Ctypansa, 271
Culicidae, 212
Cupido, 289
Curculionidae, 201–206
Curinus, 177
Curtonotidae, 230
Curtonotum, 230

Cyanallagma, 20
Cyclocephala, 162–163
Cycloglypha, 312
Cycloneda, 177
Cydnidae, 96
Cydosia, 283
Cylindera, 151
Cyphoedma, 263
Cyphomyia, 216
Cyphonia, 133
Cyrea, 178
Cyrtocapsus, 78
Cyrtonota, 190
Dalceridae, 249
Damselflies, 8-10, 17-23
Danaus, 245, 308
Darnis, 135
Dasydactylus, 174
Deloyala, 188
Delphacidae, 107–108
Deltocephalinae, 130-131
Deltochilum, 158
Denticollinae, 167
Deraeocorinae, 78
Derbidae, 105–106
Derobrachus, 149, 182
Desmia, 253
Detrivora, 243, 291
Dexosarcophaga, 232
Diabrotica, 193–195
Diacira, 111
Diacme, 251
Diactor, 92
Diaethria, 300
Diaphania, 242, 251
Diapheromeridae, 47–48
Diaphorostylus, 215
Diaphorus, 222
Diapriidae, 324
Diareusa, 110
Diastatops, 13
Dichrorampha, 247
Dicranomyia, 210
Dicromantispa, 144
Dictyopharidae, 109, 112–113
Dilobopterus, 126
Dilophus, 211
Dineutus, 154
Diogmites, 218
Dione, 294
Diparinae, 325
Diplectrus, 182
Dipropus, 166
Diptera, 207-234
Dircema, 192
Disceratus, 39
Discodon, 172
Dismorphia, 286
Disonycha, 195
Disphragis, 269
Disteniidae, 186
Dobsonflies, 143, 145
Dolichoderus, 354
Dolichopodidae, 220–222
Dolichosomastis, 270
Doxocopa, 299
Draeculacephala, 123
Dragonflies, 9-17, 24
Drapetes, 166
Drosophila, 230
Drosophilidae, 208, 230–231
Dryadula, 294
Drymophilacris, 28
Dysdercus, 74, 87–88
Dysschema, 272
Dythemis, 13–14
Dytiscidae, 154
Earwigs, 358-359
Eburia, 182
Eciton, 349
Ectatomma, 351
Ectobiidae, 64, 66–67, 69
Ectrichodiinae, 78
Edessa, 100-101
Efferia, 217
Eicherax, 218
Elachistidae, 308
Elaphidion, 184
Elateridae, 165–167
Empididae, 220
Enallagma, 21
Enchenopa, 132
Enchophora, 4, 110

Endomychidae, 178–179

Eneoptera, 36

Eneopterinae, 36–37

Enicospilus, 327

Entheus, 312

Entomopathogenic fungus,73, 149, 245

Eois, 267

Epargyreus, 309

Ephoria, 255

Epilachna, 178

Epilampra, 67

Epimecis, 262

Epiperola, 249

Epiphragma, 211

Epipsocidae, 359

Episactidae, 34

Episactus, 34

Epitausa, 271

Epopterus, 179

Epuraea, 173

Eragisa, 269

Erastria, 262

Erebidae, 244, 270–282

Eresia, 297

Erinnyis, 258

Eriopis, 177

Eristalis, 222

Erotylidae, 174–176

Erythemis, 15, 24

Erythrodiplax, 14–15

Erythrogonia, 126-127

Eubaphe, 268

Euceraia, 37

Eucereon, 275-276

Eucharitidae, 325

Euchroma, 164

Eudaminae, 314

Eufriesea, 346

Euglossa, 346

Euglyphis, 261

Eugnamptus, 200

Eulaema, 347

Eulophidae, 325

Eumaeus, 289

Eumastax, 34

Eumecosoma, 219

Eumeninae, 332

Eumolpinae, 200

Eumorpha, 259-260, 315

Euparyphus, 214

Eupastranaia, 254

Eupelmidae, 325–326

Euphorinae, 328

Euphyllodromia, 66

Euplatypus, 206

Euptoieta, 294

Euptychia, 306

Eurema, 285

Eurybia, 290

Euryneura, 215

Euschistus, 97

Eusitalces, 29

Euteliidae, 309

Euthyrhynchus, 102

Euxesta, 227

Evaniella, 324

Evaniidae, 324

Evius, 278

Exomalopsis, 345

Exophthalmus, 201

Exora, 193

Farigia, 269

Fecelia, 96

Feigeria, 281

Fibrenus, 88

Fidicina, 137

Fidicinoides,

Flatidae, 73, 108

Flatormenis, 108

Flies, 207-234

Florilegus,

Forcipomyia, 213

Forcipomyiinae, 213

Forficulidae, 360

Formicidae, 318-321, 348–354

Fountainea, 302

Froghoppers, 114-117

Fucellia, 231

Fulgoridae, 4, 70, 109–111

Fulvius, 77

Galgupha, 103

Gamelia, 256

Gelastocoridae, 75–76
Gelastocoris, 75–76
Geocoridae, 85
Geocoris, 85
Geometridae, 262–268
Gerridae, 75
Ghinallelia, 78
Gigantiops, 350
Givira, 248
Glyphidops, 226
Glyphodes, 252
Glyphonyx, 166
Goja, 360
Gomphidae, 9
Gomphocerinae, 32
Goniadera, 180
Gonodonta, 270
Gorgone, 270
Gorgythion, 313
Gorybia, 184
Grallipeza, 225
Graphocephala, 127
Grasshoppers, 27-35
Greta, 307
Griburius, 198
Gryllidae, 35–37
Gryllotalpidae, 37
Guayaquila, 319
Gynacantha, 11
Gynandrobrotica, 195
Gyrinidae, 154
Hadria, 120
Haetera, 316
Halictidae, 341–342
Hamadryas, 300-301
Hansreia, 157
Haplogleniinae, 143
Harpactorinae, 83
Haywardella, 304
Hedriodiscus, 214
Heilipodus, 205
Heilipus, 204
Heliastus, 28
Heliconius, 294–296
Heliopetes, 313
Heliscus, 155
Helocassis, 189
Hemeroblemma, 282
Hemiargus, 289
Hemipterodes, 266
Heniartes, 81
Hermetia, 216
Hermeuptychia, 306
Hesperiidae, 309–314
Hesus, 84-85
Hetaerina, 17–18
Heteracanthia, 214
Heteragrion, 23
Heteragrionidae, 23
Heteroptera, 72, 74-103
Heterosais, 307
Hexatoma, 211
Heza, 81
Hilarempis, 220
Hilarographa, 247
Hilipinus, 205
Himerarctia, 280
Hirtodeceia, 228
Hirtodrosophila, 208, 231
Histeridae, 155
Historis, 298
Holophris, 328
Homalinotus, 205
Homalodisca, 128
Homeomastax, 34
Homoeocera, 277
Homophileurus, 163
Hoplia, 157
Hoplocopturus, 203
Hoplomutilla, 329
Horiola, 133
Hortensia, 127
Hositea, 253
Hyalodictyon, 113
Hyalothyrus, 309
Hyalymenus, 89–90
Hybobathus, 223
Hybos, 220
Hybotidae, 220
Hydata, 266
Hydrophilidae, 154
Hydrophilus, 154
Hydropsychidae, 238–239

Hylephila, 312
Hylesia, 257
Hymenoptera, 317-354
Hypanthidioides, 342
Hyperandra, 273
Hypselonotus, 94-95
Ichneumonidae, 326
Ichthyobelus, 127
Idaea, 267
Idalus, 279
Idiocerinae, 117
Idiopteron, 167
Imatidium, 191
Inuyana, 127
Iphiclus, 175-176
Iphirhina, 116
Iridopsis, 262
Isanthrene, 281
Iscadia, 282
Ischnocentrus, 131
Ischnocodia, 190
Ischnura, 10, 21
Ischyrus, 174
Isodontia, 338
Isonychus, 159
Issidae, 113–114
Ithomia, 307
Jadera, 96
Jakrama, 128
Jalysus, 85
Jobertus, 78
Joppa, 326
Junonia, 299
Katydids, 25-26, 37-40
Keroplatidae, 212
Kuschelina, 195
Laccogrypota, 114
Lacewings, 140,
Ladoffa, 128
Lagria, 180
Lamiinae, 184-186
Lamprocera, 170
Lamprosphaerus, 199
Lampyridae, 168–170
Lappida, 112
Largidae, 88–89
Largus, 89
Lasaia, 290
Lasiocampidae, 242, 261
Lasioglossum, 341
Lasiophanes, 350
Latochrimnus, 86
Lauxaniidae, 227–229
Leafhoppers, 117-131
Leiopleura, 164
Leistotrophus, 156
Lema, 186
Leptinotarsa, 199
Leptoceridae, 240
Leptoglossus, 92
Leptomerinthoprora, 30
Leptonema, 238
Leptophlebiidae, 359
Leptophobia, 286
Leptoscelis, 93
Leptostales, 267
Leptostylopsis, 184
Leptotes, 289
Leptotettix, 39
Leptysminae, 28
Lerema, 310
Lestes, 23
Lestidae, 23
Letis, 281–282
Leucanella, 257
Leuciris, 262
Leucotmemis, 278
Libellula, 16
Libellulidae, 11–17
Ligyra, 216
Limacodidae, 249, 314
Limnichidae, 165
Limnichites, 165
Limnogonus, 75
Limnophilinae, 211
Limoniidae, 210–211
Lintneria, 259
Lissoteles, 219
Lithophorus, 176
Lithoscirtus, 28
Liturgusa, 54-56
Liturgusidae, 54–56
Lixodes, 205

Lobetus, 171
Lobopoda, 179
Lochmorhynchus, 217
Lophocampa, 273
Lophoscutus, 84
Loxa, 97
Loxophlebia, 278
Lucidota, 169-170
Lycaenidae, 287–289
Lycambes, 91
Lycidae, 167–168
Lycoderes, 134
Lycoderides, 134
Lyctinae, 172
Lygaeidae, 86
Lygaeus, 86
Lymexylidae, 172
Lystra, 70
Macaria, 263
Machilidae, 358
Macrocneme, 276
Macrocondyla, 216
Macrocopturus, 204
Macrodactylus, 158–159
Macromusonia, 60
Macronema, 239
Macrostemum, 239
Macrostola, 173
Macrothemis, 17
Macunolla, 128
Madoryx, 259
Magava, 268
Mahanarva, 117
Malacomorpha, 42, 45
Mallophora, 217
Malocampa, 269
Manduca, 258
Mantidae, 56–60
Mantises, 49-62
Mantisflies, 144
Mantispidae, 144
Mantoida, 52
Mantoididae, 52
Mapeta, 250
Marilia, 240
Marmarina, 160
Maronius, 171
Marpesia, 299-300
Marshallius, 205
Maruca, 253
Mastixis, 272
Maxantonia, 115
Mayflies, 358
Mechanitis, 308
Mecistogaster, 22
Medetera, 220
Megachilidae, 342
Megaloprepus, 22
Megalopygidae, 249–250
Meganola, 282
Megapodagrion, 23
Megapodagrionidae, 23
Megaselia, 222
Megasoma, 163
Megathecla, 287–288
Megophthalminae, 117
Melanis, 293
Melanoliarus, 104
Melanomyza, 227
Melete, 286
Melipona, 344-345
Melittia, 248
Melucha, 94
Melyridae, 172
Membracidae, 131–136, 321
Membracis, 135-136, 321
Memphis, 302
Merosargus, 215
Mesopteron, 167
Mesosemia, 290
Mestra, 301
Metacanthus, 85
Metacycla, 193
Metamasius, 202–203
Metapelma, 326
Metapteron, 168
Metaxyonycha, 200
Metilia, 52–53
Metria, 282
Micrathyria, 12
Micrauchenus, 81
Microbembex, 339
Microcentrum, 37

Microcerella, 232
Microchrysa, 215
Microctenochira, 189
Microctonus, 328
Microgastrinae, 328
Micropeza, 226
Micropezidae, 225–226
Microsepsis, 230
Microstigma, 22
Mictopsichia, 246
Midila, 253
Mimallonidae, 254
Minettia, 228
Minthea, 172
Minytus, 76
Miresa, 249
Miridae, 76–78
Misagria, 11
Mischocyttarus, 334-335
Mocis, 282
Monecphora, 115
Monocesta, 198
Monophlebidae, 138
Montina, 80
Mordellidae, 181
Mormidea, 97
Morpheis, 248
Morpho, 241, 307
Moths, 241-283
Muirolonia, 105
Murgantia, 98
Musca, 231
Muscidae, 231–232
Musonia, 61
Mutillidae, 329
Mycotretus, 174-175
Mydromera, 276
Mylaris, 179
Myodocha, 87
Myrmeleontidae, 141, 144
Myrmicinae, 318, 352-354
Mysidia, 106
Myzinum, 329
Nannotrigona, 343
Nasutitermes, 64–65
Nasutitermitinae, 64–65, 68
Naupactus, 202
Neanastatinae, 325
Nectopsyche, 240
Neivamyrmex, 349
Nematocampa, 262
Nematocera, 208, 210-211
Nematodes, 321
Nematopus, 95
Nemobiinae, 36
Nemorhinus, 201
Nemoria, 265–266
Neoconocephalus, 40
Neocurtilla, 37
Neoerythromma, 20
Neogriphoneura, 228
Neohalyzia, 177
Neolema, 186-187
Neomalaxa, 107
Neomegalotomus, 90
Neoneura, 21
Neopamera, 87
Neoponera, 348
Neoprocis, 250
Neoptychodes, 185
Neotynelia, 133
Nephrotoma, 210
Neriidae, 226
Neritos, 274
Nersia, 113
Neruda, 296
Neuraeschna, 11
Nialaphodius, 156
Nica, 301
Nicomiinae, 132
Nilio, 180
Nitidulidae, 173
Noctuidae, 283
Nogodinidae, 111–112
Nolidae, 282
Nothomyia, 214
Notiobia, 153
Notocyrtus, 80
Notodontidae, 268–269
Notofenusa, 323
Notonecta, 75
Notonectidae, 75
Nyceryx, 258

Nyctelia, 181
Nymphalidae, 293–308, 314, 316
Nymphidium, 293
Nymphula, 252
Ocalaria, 272
Ocoaxo, 115
Octotoma, 188
Ocyptamus, 223-224
Odontoceridae, 240
Odontocheila, 152
Odontomachus, 348
Oecetis, 240
Oecleus, 105
Oedancala, 86
Oedemeridae, 181–182
Oedipodinae, 28-29
Olceclostera, 255
Oleria, 308
Oligoclada, 12
Oligocorynus, 176
Oligopleura, 268
Oligotomidae, 360
Omalotettix, 29
Omicron, 332
Omiodes, 252
Omiostola, 247
Ommatolampidinae, 29-30
Ommatolampis, 30
Omolicna, 106
Omophoita, 196
Omura, 35
Oncometopia, 129
Oncopeltus, 86
Oncotophasma, 47-48
Onega, 121
Onicideres, 186
Ontherus, 158
Onychocerus, 184
Oospila, 264-266
Ophoelis, 170
Opisthoxia, 263
Oragua, 121
Orasema, 325
Oraseminae, 325
Oreodera, 185
Orfeliini, 212
Ormenoides, 108
Ormetica, 280
Ornidia, 222
Orophus, 38
Orphulella, 31
Orthemis, 16
Orthomegas, 182
Orthorene, 256
Ortilia, 297
Osbornellus, 130
Ostha, 270
Owlflies, 143, 146
Oxelytrum, 155
Oxyagrion, 21
Oxycopis, 181
Oxyopsis, 57
Oxysternon, 158
Oxytenis, 256
Ozophora, 87
Pachybrachis, 198
Pachycondyla, 348
Pachycoris, 103
Pachygastrinae, 215
Pachygronthinae, 86
Pachylia, 258-259
Pachylis, 72, 95
Paectes, 309
Paederus, 155
Palaemnema, 23
Palingonalia, 124
Palpada, 223
Palpibracus, 232
Panchlora, 68
Pangaeus, 96
Papias, 310
Papilio, 284
Papilionidae, 284, 314
Parachartergus, 322, 336
Paraclius, 220
Paracometes, 186
Paracrocampsa, 129
Paragonia, 263
Paragrallomyia, 225
Paralycambes, 91
Paramasius, 203
Paranapiacaba, 193
Paraponera, 349

Parasa, 249
Parasitalces, 30
Parastagmatoptera, 59
Parastratocles, 46
Parathona, 121
Paratrechina, 71
Paratriarius, 196
Paraulacizes, 129
Pareuptychia, 306
Parevia, 278
Parides, 284, 314
Parimatidium, 192
Paromenia, 122
Partamona, 343, 345
Paryphes, 93
Passalidae, 155
Passalus, 155
Paulogramma, 300
Peiratinae, 84
Pelidnocoris, 99
Pelidnota, 161
Pellaea, 98
Pelmatosilpha, 66
Peltidocylapus, 78
Pemphredoninae, 340
Penniverpa, 220
Pentacomia, 152
Pentagonica, 153
Pentatomidae, 72, 96–102
Pentobesa, 268
Pephysena, 87
Pepsis, 330-331
Peranthus, 91
Perichares, 311
Peridinetus, 206
Periphoba, 257
Periplaneta, 63, 66
Periscelididae, 230
Perithemis, 16
Perlidae, 359
Perola, 249
Peromatus, 102
Perophthalma, 291
Petalops, 93
Petrophila, 252
Phaenolis, 170
Phaeomolis, 276
Phalangopsidae, 37
Phaloe, 273
Phanaeus, 158
Phaneropterinae, 38
Phasiinae, 233
Pheidole, 318, 353
Philaethria, 296
Philophyllia, 38
Phocides, 310
Phoebis, 285
Phoridae, 222
Photinus, 168-169
Phrenapates, 181
Phrictopyga, 107
Phrictus, 110
Phygadeuontinae, 327
Phylloptera, 38
Phytocoris, 76-77
Piazurus, 203
Picumna, 113
Pierella, 304-305
Pieridae, 242, 285–287
Piezogaster, 96
Pilicata, 219
Pintalia, 104
Placogenys, 85
Plagiogramma, 155
Plagiometriona, 190
Plagioneurus, 221
Planicephalus, 130
Planthoppers, 103-114
Platerodinae, 167
Plateros, 167
Platygastridae, 324–325
Platygonia, 122
Platyphora, 198–199
Platystictidae, 23
Plectoderes, 109
Plectrophoroides, 202
Plesiochrysa, 142
Plocoscelus, 225
Poanes, 311
Poblicia, 111
Podagrion, 50
Podisus, 102
Podoscirtinae, 35-36

Poecilominettia, 229
Poecilopompilus, 331
Poecilopsocus, 360
Poecilotylus, 226
Poekilloptera, 108
Polistes, 332–334
Polistinae, 318, 322, 332-337
Polybia, 322, 336-337
Polyglypta, 133-134
Polymerus, 77
Polythrix, 314
Pompilidae, 330–331
Ponana, 118
Potnia, 132
Prasinochrysa, 267
Prepops, 77
Priocnemella, 331
Priocnessus, 331
Prioninae, 149, 182
Prionolopha, 32
Prisopus, 45
Proconiini, 128-129
Proctolabinae, 28
Progomphus, 9
Prolatiforceps, 218
Prolepsis, 234
Prometopia, 173
Prosapia, 115
Proscopia, 34
Proscopiidae, 34
Proselandria, 323
Prosopantrum, 230
Protalebrella, 131
Prothysana, 314
Protodictya, 229
Protopolybia, 337
Proxys, 97
Pselaphidae, 156
Pselliopus, 79
Pseudococcidae, 71
Pseudodirphia, 257-258
Pseudomethoca, 329
Pseudomiopteryginae, 62
Pseudomiopteryx, 62
Pseudomops, 67
Pseudomyrmex, 351
Pseudopanurgus, 341
Pseudophasma, 44
Pseudophasmatidae, 42, 44–46
Pseudophyllinae, 39
Pseudoxycheila, 151
Psocidae, 359
Psocoptera, 356, 359
Psorophora, 212
Psychocampa, 254
Psychodidae, 212
Psyllidae, 138
Psyllobora, 178
Ptecticus, 215
Pterodictya, 111
Pteromalidae, 325
Pterophoridae, 246
Ptilodactyla, 165
Ptilodactylidae, 165
Ptiloglossa, 340
Ptilosphen, 225
Pycnarmon, 252
Pycnopalpa, 39
Pygodasis, 330
Pyractonema, 169
Pyralidae, 250
Pyrgomorphidae, 35
Pyrgotidae, 227
Pyrophorus, 165 P
Pyrrhocoridae, 74, 87–88
Pyrrhogyra, 301-302
Pyrrhopyge, 313
Quentalia, 255
Raphirhinus, 129
Raphirhynchus, 201
Rasahus, 84
Reduviidae, 78–84
Remella, 311
Repipta, 82
Rhabdatomis, 274
Rhagovelia, 75
Rhescyntis, 255
Rhiginia, 78, 83
Rhinostomus, 202
Rhipha, 274
Rhodochlora, 265
Rhodopygia, 15
Rhopalidae, 96

Rhopalosoma, 328
Rhopalosomatidae, 328
Rhynchitidae, 200
Rhynencina, 227
Rhyparochromidae, 87
Rhytidochrota, 31
Richardia, 226
Richardiidae, 226
Ricolla, 82
Riodina, 291
Riodinidae, 243, 290–293
Ripipterygidae, 35
Ripipteryx, 35
Roelofa, 254
Rogadinae, 328
Romaleidae, 32–33
Ropalomeridae, 210, 229
Rosema, 269
Rutela, 161
Sagotylus, 94
Sarcophagidae, 232
Saturniidae, 255–258, 315
Satyrotaygetis, 305
Sawflies, 323
Scale Insects, 71, 138
Scaphytopius, 131
Scarabaeidae, 150, 156–164
Scaris, 118
Scarites, 153
Scathophaga, 231
Scathophagidae, 231
Scelioninae, 325
Sceliphrinae, 338
Sceliphron, 338
Sceloenopla, 188
Schistocerca, 32
Schraderiellus, 99
Sciomyzidae, 229
Sciponacris, 30
Scoliidae, 330
Scolytinae, 206
Scopogonalia, 122
Scotobius, 181
Scudderia, 38
Scutelleridae, 103
Selandriinae, 323
Selenophorus, 153-154
Selymbria, 136
Semiotus, 166
Semnia, 250
Sepsidae, 230
Sericoptera, 264
Sesiidae, 248
Setabis, 292
Sevia, 109
Sharpshooters, *see Leafhoppers*
Sibaria, 98
Sibovia, 122-123
Silinae, 172
Silis, 172
Silphidae, 155
Silvanidae, 173
Silvanus, 173
Simena, 264
Simuliidae, 212
Simulium, 212
Sinopia, 115
Siphonophysa, 229
Siproeta, 298-299
Skippers, 309-314
Smiliinae, 133-134
Sogatella, 107
Soosiulus, 123-124
Sosxetra, 283
Sphacelodes, 263
Sphecidae, 338–339
Sphecinae, 338
Sphecomyiella, 227
Sphenorhina, 116
Sphex, 338
Sphingidae, 258–261, 315
Spiders, 209, 318
Spintherophyta, 200
Spittlebugs, 114-117
Spongiphoridae, 359
Spongophora, 359
Stagmomantis, 58
Stalachtis, 291
Staphylinidae, 155–156
Statira, 180
Stegaspis, 134
Stehlikiana, 123
Stelidota, 173

Stenacris, 28
Stenocoris, 91
Stenodorus, 35
Stenodynerus, 332
Stenoma, 308
Stenomacra, 89
Stenomicra, 230
Stenophyllinae, 53
Stenopoda, 83
Stenoscelidea, 94
Stenosigma, 332
Stenotarsus, 178
Stephanidae, 323
Stephanolla, 123
Sternorrhyncha, 71, 137-138
Sternurus, 200
Sterphus, 223
Stick Insects, 42-48
Stilpnochlora, 39
Stirellus, 130
Stolas, 190
Stoneflies, 358
Strategus, 164
Stratiomyidae, 214–216
Stratocleinae, 46
Strigoderma, 161
Strongylium, 179
Strophocerus, 268
Strymon, 287–288
Stylocentrus, 134
Syllepis, 253
Symphata, 323
Symphylus, 102–103
Synchlora, 266
Syngamia, 253
Synoeca, 322, 337
Synoestropsis, 239
Syntermes, 68
Syntomacris, 30
Syntomeida, 277
Syrphidae, 222–224
Systena, 196
Tabanidae, 214
Tachinidae, 232–233
Taedia, 77
Taeniogonalos, 324
Taeniopoda, 32-33
Taeniotes, 185
Tafalisca, 36
Talanus, 179
Talides, 311
Tamyra, 250
Tanypus, 214
Taosa, 113
Taperinha, 130
Taphura, 136
Tatochila, 286
Taygetis, 305
Tegosa, 297
Telchin, 248
Telebasis, 22
Telephanus, 173
Temenis, 302
Tenaspis, 170
Tenebrionidae, 179–181
Tenthredinidae, 323
Tephritidae, 227
Termites, 64-65, 68
Termitidae, 64-65, 68
Tethlimmena, 183
Tetracha, 151
Tetragona, 345
Tetrataenia, 28
Tetreuaresta, 227
Tetrigidae, 35
Tettigoniidae, 37–40
Tettisama, 126
Teucholabis, 211
Thasus, 72
Thaumatographa, 247
Theages, 275
Thecomyia, 229
Theope, 290
Theraneis, 89
Therevidae, 220
Therinia, 256
Thespidae, 60–62
Thionia, 114
Thrinaconyx, 61
Thrips, 359
Thynnidae, 329
Timulla, 329
Tingidae, 84

Tipulidae, 210
Titya, 242
Tmetomorpha, 263
Tmolus, 288
Tolania, 132
Tolypida, 254
Tortricidae, 246–247
Tortyra, 246
Torymidae, 50
Toxomerus, 224
Toxotarsinae, 232
Trachyderes, 184
Trachymyrmex, 353
Treehoppers, 131–136, 321
Trentepohlia, 208-211
Triacanthagyna, 11
Trichobrotica, 197
Trichophora, 232
Trichopria, 324
Trichoptera, 237-240
Trichoscelia, 144
Trichostictia, 339
Trichoxys, 148, 183
Trichromia, 280-281
Trigona, 319, 343–344
Trigonalidae, 324
Trigonidiidae, 36
Trimerotropis, 29
Tropicanus, 131
Tropidacris, 27, 33
Trosia, 250
Trypoxylon, 340
Tubulifera, 359
Tuina, 274
Turuptiana, 279
Tylozygus, 124
Ugyops, 107-108
Ulidiidae, 227
Uracis, 12–13
Urania, 261
Uraniidae, 261
Urbanus, 309-310
Utetheisa, 273
Vanessa, 297
Vates, 56
Vatinae, 56-57
Vehilius, 311
Veliidae, 75
Vespidae, 332–337
Vestistilus, 134
Vestria, 40
Vettius, 312
Villa, 217
Vutina, 112
Waldheimia, 323
Walkingsticks, 42-48
Walterianella, 195
Wasps, 317-318, 322-340
Webspinners, 359
Weevils, 200-206
Willistoniella, 210, 229
Xantholinini, 156
Xanthopimpla, 327
Xenosoma, 276
Xerosomatinae, 46
Xestocephalus, 117
Xiphiola, 30
Xyleborus, 206
Xyleus, 32
Xylocopa, 343
Xylophanes, 260, 315
Xylospinodes, 46
Xylotribus, 186
Zaprionus, 231
Zatrephina, 192
Zelomorpha, 328
Zelurus, 79
Zelus, 82–83
Zeugomantispa, 144
Zoolea, 57
Zoreva, 94
Zulia, 117

Made in the USA
Las Vegas, NV
03 November 2026

57422975R00226